T0009240

"In a world where even our dreams can be mined for data, we need more guides like [Farahany] to think through the challenges ahead."
—*The Wall Street Journal*

WITH A NEW AFTERWORD BY THE AUTHOR

The Battle for Your Brain

DEFENDING THE RIGHT TO THINK FREELY IN THE AGE OF NEUROTECHNOLOGY

Nita A. Farahany

Praise for

The Battle for Your Brain

"Farahany convinced me that the reality of brain-tracking is much closer than I imagined." —Jessica Grose, *The New York Times*

"The brain is the new battleground for humanity. And regardless of what governments do or don't do to regulate technologies or protect citizens, it will be the responsibility of each and every one of us to now think about how we are going to protect our own brains."

—*Forbes*

"*The Battle for Your Brain* is a superb introduction to how rapidly advancing neurotech can either enhance or undermine free minds."

—*Reason*

"The author's evenhanded approach is a refreshing reprieve from the dystopian pessimism that often accompanies discussions of these technologies, and the eye-popping examples show that the future may be closer than many assume. Readers will be enthralled."

—*Publishers Weekly*

"An unsettling warning . . . an insightful report." —*Kirkus Reviews*

"An essential guide to a brave new world where your thoughts are no longer private. In part a cautionary tale, in part an episode of *Westworld*, this is the first of many books to be written on the subject. Expect no other author to match Farahany's intellectual grit. She deserves an audience at our highest policy-making institutions."

—Scott Galloway, professor of marketing at New York University's Stern School of Business and bestselling author of *Adrift*

"Few know better than Farahany what neurotech can do for people's betterment, and this lends weight to *The Battle for Your Brain*. Her book is a somber, troubling, yet essential account of a field whose speed of expansion alone should give us pause." —Simon Ings, *New Scientist*

"Smart and thought-provoking. Ask yourself: What will it mean if our thoughts and emotions are up for grabs, just like the rest of our data?"

—Gintaras Radauskas, *Cybernews*

"Nita Farahany writes with clarity and verve about the promise and perils of the neurotech revolution—offering a fascinating and provocative tour of technologies that have the power to transform our lives for the better and even what it means to be human. More importantly, she encourages a timely global conversation about how to ensure the ethical progress of neurotech to benefit all of humanity."

—Jennifer Doudna, biochemist, University of California, Berkeley; Innovative Genomics Institute founder; and Nobel Laureate for coinventing CRISPR technology

"Essential reading for anyone interested in neurotechnology and its coming impact on our society. Engineering neural implants to decode the brain seems hard to fathom, but this is easy compared to the ethical challenges that lie ahead. Farahany masterfully navigates the issues that confront us."

—Edward Chang, MD; chairman of the University of California, San Francisco's Department of Neurological Surgery

"Farahany poses the critical questions that can guide us as we navigate the hope and hype around neurotechnology, revealing both the promise for patients and the challenge for society. *The Battle for Your Brain* is a must-read."

—Thomas Insel, MD; author of *Healing*; former National Institute of Mental Health director; and codirector of President Barack Obama's BRAIN Initiative

"This highly original and timely book explains why we cannot surrender our 'last bastion of freedom,' even as we fight with politics and persuasion for access to the fruits of brain science. Farahany alerts us to a struggle for control over access to sensitive personal information that demands everyone's attention."

—Anita Allen, Henry R. Silverman Professor of Law and professor of philosophy, University of Pennsylvania Carey Law School

"Farahany sounds a timely warning concerning current uses of neurotechnology by corporations and governments for monitoring, recognizing that these uses will grow more powerful and insidious. However, she is no enemy of technology. She presents a balanced view of risks and benefits of its uses by individuals, and makes her arguments in the context of a sophisticated understanding of individual liberty and its potential limits in a free society."

—Steven Hyman, MD; director of the Stanley Center for Psychiatric Research; and member of the Broad Institute of MIT and Harvard

"Nita Farahany persuasively demonstrates that rapidly approaching advances in neurotechnology will change politics, marketing, mental-health care, and dozens of other areas of daily life. The legal and ethical challenges she outlines are daunting, but *The Battle for Your Brain* arms us with the knowledge needed to fight for a future that includes individual privacy and free will."

—Jules Polonetsky, CEO, Future of Privacy Forum

"As a well-established thought leader in ethics and artificial intelligence, Professor Farahany is neither [an] alarmist nor [is she] resigned over current trends, but she offers a measured manifesto of how we can channel technological progress for the benefit of humanity. However, the message is clear: if we do not institute the necessary safeguards now, humanity as we know it is imperiled."

—Ahmed Shaheed, professor, University of Essex, and author of the first-ever report on freedom of thought for the United Nations

The Battle for Your Brain

DEFENDING THE RIGHT TO THINK FREELY IN THE AGE OF NEUROTECHNOLOGY

Nita A. Farahany

ST. MARTIN'S GRIFFIN
NEW YORK

Published in the United States by St. Martin's Griffin, an imprint of St. Martin's Publishing Group

For information, address St. Martin's Publishing Group,
120 Broadway, New York, NY 10271.

www.stmartins.com

Design by Meryl Sussman Levavi

The Library of Congress has cataloged the hardcover edition as follows:

Names: Farahany, Nita A., author.
Title: The battle for your brain : defending the right to think freely in the age of neurotechnology / Nita A. Farahany.
Description: First edition. | New York : St. Martin's Press, 2023. Includes bibliographical references and index.
Identifiers: LCCN 2022051602 | ISBN 9781250272959 (hardcover) | ISBN 9781250272966 (ebook)
Subjects: LCSH: Neurosciences—Moral and ethical aspects—Popular works. | Neurotechnology (Bioengineering)—Moral and ethical aspects—Popular works.
Classification: LCC RC343 .F37 2023 | DDC 616.8—dc23/eng/20221202
LC record available at https://lccn.loc.gov/2022051602

ISBN 978-1-250-33927-0 (trade paperback)

First St. Martin's Griffin Edition: 2024

1 3 5 7 9 10 8 6 4 2

To Mom and Dad,

for always believing in me, even when they think

that I have no idea what I'm talking about

Contents

Introduction

I'm trying to get the birds to sing. If I can calm my mind just enough, they will *sing.*

Though I've tried to convince my children otherwise, I don't have magical powers. I'm wearing a simple headband embedded with electrodes that detect my brain wave activity and send it via Bluetooth to an application on my iPhone. Brain waves—the oscillating electrical voltages in your brain—are small in size (just a few millionths of a volt), but they reveal a lot about the inner workings of your mind. When I relax into a meditative state, my alpha brain wave activity rises, and the app rewards me with the sound of singing birds. This neurofeedback technique has proven powerful in preventing the migraine attacks that have dogged me since childhood.

Having used the device a number of times, I know how to increase my alpha waves—the pattern of electrical activity produced by the brain when you're feeling calm and peaceful—and reduce my beta wave activity—the higher-speed brain waves that occur when you're wide awake and thinking. I focus on a happy memory:

My eldest daughter, Aristella, is just three years old, and we are hiking as a family to a waterfall in the mountains of North Carolina. The season is late fall; afternoon sun glistens through trees still dotted with red and orange as we crunch through the fallen leaves beneath our feet. I can almost feel the sun on my face and hear the water gurgling over waterworn rocks. In my mind, I hear Aristella's peals of laughter as she races a leaf downstream against one tossed in by my husband, Thede.

Chirp, chirp! The app confirms that my brain is responding.

Whether we are meditating, doing a math calculation, recalling a phone number, or browsing through our mental thesaurus for just the right word, neurons are firing in our brains, creating minuscule electrical discharges. When a mental state like relaxation or stress is dominant, hundreds and thousands of neurons are firing in characteristic patterns that can be measured with an electroencephalogram, or EEG.

Scientists used to have to place electrodes directly on the periosteum—the inner layer of the scalp—to pick up brain waves. The procedure required surgery under anesthesia and carried risks, including fever, infection, and leaking brain fluid. Today, the electrodes can be placed externally, on the forehead or the surface of the scalp. EEG devices detect and record brain waves in terms of cycles per second, known as hertz (Hz).[1] Alpha waves, for example, clock in at the 8–13 Hz range.

Had I wanted to, I also could have placed electrodes over the muscles on my body, to measure the signals sent to them while I was meditating. Our brains are constantly transmitting signals to our peripheral nervous system—the parts of the nervous system beyond the brain and the spinal cord. Electromyography (EMG) can be used to detect the electrical activity in response to a nerve's stimulation of the muscle in millivolts, ranging from 0 to 10 mV (+5 to −5).[2] Together, EEG and EMG give us a window on what our brain is up to at any given moment, including the instructions it is sending to the rest of the body.

Our use of EEG and EMG draws on discoveries made by two

Italian scientists in the late 1700s regarding the electric battery and bioelectrical activity in the body. More recent technological leaps in neuroscience and artificial intelligence have converged to give us consumer neurotech devices—a catchall term for gadgets that connect human brains to computers, and the ever more sophisticated algorithms that allow those computers to analyze the data they receive. At first, neuroscientists rightly dismissed all these consumer devices as inaccurate and unvalidated, little better than toys. But as both the hardware and software improved, consumer neurotech became more accurate and harder to dismiss. The average tech-savvy person can now "see" their emotions,[3] arousal, and alertness,[4] and track how effectively they are meditating.[5]

Personal neurotech devices are just one part of the growing category of wearable tech, which allows the average technophile to quantify their bodily functions. The category is so popular that as of 2020, nearly one out of every five Americans was using one.[6] There are more than three hundred thousand different mobile health apps available worldwide (a number that has doubled in just five years), with an estimated market value surpassing $100 billion.[7]

Globally, the market for neurotechnology is growing at a compounded annual rate of 12 percent and is expected to reach $21 billion by 2026.[8] Consumers can see graphic displays of their brain wave activity in real time—delta (dreamless sleep), theta (deep relaxation, daydreaming, inwardly focused), alpha (very relaxed, taking a break, meditating), beta (aroused, engaged, stressed), and gamma (concentrating) waves—as well as patterns of blood flow in their brains and even the bioelectric changes in their muscles.[9]

Self-tracking is far more than a fad. It's a new way of living and thinking about ourselves.

Neurotechnology can tell us if we're wired to be conservative or liberal,[10] whether our insomnia is as bad as we think,[11] and if we're in love with someone or just "in lust."[12] We can learn how we process risks and rewards and whether we're congenitally disposed to be spendthrifts or tightwads.[13] Soon, smart football helmets will be able to diagnose concussions immediately after they occur. Neurotech devices can also track changes in our brains over time, such as the slowing down of activities in certain brain regions associated with the onset of conditions like Alzheimer's disease,[14] schizophrenia,[15] and dementia.[16] Not everyone wants to know if one of those conditions is in the cards for them, but those who do may benefit from having time to prepare.

Neurotech companies are already marketing technology to detect drivers who may be drowsy and prevent them from falling asleep at the wheel. A simple, wearable device that measures EEG can alert individuals who have epilepsy to oncoming seizures, while those with quadriplegia can type on computers using just their thoughts. I am excited by the promise of this technology to help people lead better lives, and a huge proponent of empowering people to take charge of their own health and well-being by giving them access to information about themselves. But there is another side to this technology, a Pandora's box that keeps me up at night.

The same neuroscience that gives us intimate access to ourselves can allow companies, governments, and all kinds of actors who don't necessarily have our best interests in mind access too. I find this terrifying as an Iranian American because nothing in the US Constitution, state and federal laws, or international treaties gives individuals even rudimentary sovereignty over their own brains. It's not going to happen tomorrow, but we are rapidly heading toward a world of brain transparency, in which scientists, doctors, governments, and companies may peer into our brains and minds at will. And I worry that in this rapidly approaching future, we will voluntarily or involuntarily surrender our last bastion of freedom: our mental privacy. That we will trade access to our brain activity to commercial entities for rebates, discounts on insurance, free access to social media accounts . . . or even as a condition for keeping our jobs.

It's already begun. In China, train drivers on the Beijing–Shanghai high-speed rail line, the busiest in the world, must wear EEG devices while they are behind the controls, to ensure that they are focused and alert. According to some news sources, workers in government-controlled factories are required to wear EEG sensors to monitor their productivity and their emotional states, and they can be sent home based on what their brains reveal.

Here in the United States, we are into the second decade of a large-scale, federally funded project called Brain Research Through Advancing Innovative Neurotechnologies (BRAIN) "aimed at revolutionizing our understanding of the human brain."[17] In 2001, DARPA (the US Defense Department's Defense Advanced Research Projects Agency) launched its hotly contested Augmented Cognition program, which develops "technologies to augment warfighters' cognitive capacity and capabilities"; its Human Assisted Neural Devices program aims to decode the human brain to better characterize and mitigate threats to it; and Active Authentication hopes to find each person's unique "cognitive fingerprint" so they can be biometrically identified. And militaries across the world are developing weapons to precisely attack our brains, "for determining victory or defeat in the battlefield."[18]

Corporations are getting in on the act too. Before long, computers will interface with our brains directly, allowing companies to know what products we want before we do—and which pitches we are the most primed to love. L'Oréal, the beauty and fragrance world leader, has even launched a strategic partnership with Emotiv, a large neurotech company, to target fragrance selection to individual brains. It now offers in-store consultations to help consumers find the "perfect scent suited to their emotions" by asking the customer to wear a multi-sensor EEG-based headset to detect and decode their brain activity through powerful machine learning algorithms.[19] Will people willingly trade their brain data for customized perfume? Will those of us who refuse the technology have advertisements targeted to our brains based on the data amassed from other people? And is this just the tip of the iceberg of corporate collection and commodification of our brains?

* * *

As an ethicist, lawyer, and philosopher, I believe that we can and should embrace emerging neurotechnology, but only if we first update our concept of liberty to maximize the benefits and minimize the risks of doing so.

Modern liberalism is undergirded by the concept of liberty given to us by the English philosopher John Stuart Mill in his 1859 classic, *On Liberty.* In the face of law or social pressure, Mill argued, individuals should have free rein in their opinions and behaviors, unless those behaviors harm other people. Mill's goal was to make happiness available as widely as possible, supporting, as he put it in his autobiography, "the importance, to man and society . . . , of giving full freedom to human nature to expand itself in innumerable and conflicting directions."[20]

We now inhabit a world Mill never imagined, in which human nature can be expanded—or restricted—by emerging neurotechnology. Imagine yourself in this scenario, in a future that is closer than many of us realize:

> *You're in the zone; even you can't believe how productive you've been. Your memo is finished and sent, your in-box is under control, and you're feeling sharper than you have in a decade. Sensing your joy, your playlist shifts to your favorite song, sending chills up your spine as the music begins to play. You glance at the program running in the background on your computer screen and notice a now-familiar sight that appears whenever you're overloaded with pleasure: your theta brain wave activity decreasing in the right central and right temporal regions of your brain.*
>
> *You mentally move the cursor to the left and scroll through your brain data over the past few hours. You can see your stress levels rising as the deadline to finish your memo approached, causing your beta brain wave activity to peak right before an alert popped up, telling you to* TAKE A BRAIN BREAK. *But what's that unusual change in your brain activity when you're asleep? It started earlier in the month. You compose a text to your doctor in your mind and send it with a mental*

swipe of your cursor: "Could you take a quick look at my brain data? Anything to worry about?"

Your mind starts to wander to the new colleague on your team, whom you know *you shouldn't be daydreaming about, given the company's policy on intra-office romance. But you can't help fantasizing just a little. You let yourself enjoy the moment and wonder if his brain aligns well with yours. Then you start to worry that your boss will notice your amorous feelings when she checks your brain activity, and shift your attention back to the present. You breathe a sigh of relief when the email she sends you later that day congratulates you on your brain metrics from the past quarter, which have earned you another performance bonus. You head home, jamming to music, with your work-issued brain-sensing earbuds still in.*

When you arrive at work the next day, a somber cloud has fallen over the office. Along with emails, text messages, and GPS location data, the government has subpoenaed every employee's brain wave data from the past year. They have compelling evidence that one of your coworkers has committed massive wire fraud; now they are looking for his coconspirators. Your boss tells you they are looking for synchronized rhythms of brain activity between him and the people he has been working with. While you know you're innocent of any crime, you've been secretly working with him on a new start-up venture. Shaking, you remove your earbuds and try to focus without them. You're not sure you remember how.

Advances in neuroscience are taking us closer to a reality like this one, where individuals, companies, and government can hack and track our brains in ways that fundamentally impact our freedom to understand, shape, and define ourselves. Which confronts us with an unprecedented set of bioethical dilemmas:

- Should we—or will we want to—directly track information gleaned from our own brains, or have that information filtered for us by a trusted intermediary instead, much as we count on medical doctors to make sense of an MRI or CT scan?

- Is it cheating to enhance our brains with drugs or devices that help us to learn faster or concentrate longer than we otherwise could?
- Does society have the right to prohibit us from slowing down our brains or extinguishing painful memories?
- What will it mean if our thoughts and emotions are up for grabs, just like the rest of our data being commodified and sold by corporations?
- Should employers be allowed to use that data as part of the growing trend of workplace surveillance?
- Are there any limits to corporations targeting our brains with their products?
- Does freedom of thought protect us from government tracking our brains and mental processes?
- Will unlocking our brains open our minds to targeted assaults and hacking, and if so, how do we protect ourselves against that risk?
- Is embracing neurotechnology necessary for the very survival of our species to compete against the growing capabilities of artificial intelligence?

This book navigates each of these dilemmas and more to help us expand our definition of liberty in the modern era to include our right to cognitive liberty—the right to self-determination over our brains and mental experiences.

Anyone who values their ability to have private thoughts and ruminations—an "inner world"—should care about cognitive liberty.

We are at a pivotal moment in human history, in which control of our brains can be enhanced or lost. We need to define the contours of cognitive liberty now or risk being too late to do so.

This book will give us the tools necessary to unpack a new right to cognitive liberty and the bundle of rights it includes—mental privacy, freedom of thought, and self-determination—while making accessible the exciting and often startling neuroscience of *tracking* and *hacking* the human brain: From how neuroscience is democratizing meditation to the coming age of neural interface and what that means to technologies that let us type on virtual keyboards and navigate virtual reality intuitively with our minds, and even detect a deadly glioblastoma before it spreads. New ways to help us focus, authenticate our identities, or interrogate our brains for crimes. How marketing companies are decoding our brains to sell us products and even tailor movies to our brains' greatest desires. Drugs and devices that can make us work and learn smarter and faster, cure us of addiction and depression, maybe even alleviate human suffering. And how our brains are manipulated and even assaulted.

But these chapters also go further, putting the science in context—demonstrating why public understanding and dialogue about cognitive liberty is more important now than at any other point in history.

My interest in cognitive liberty goes to the very heart of who I am. It affects you just as deeply, but so far, the issue hasn't sparked nearly so much concern as I believe it should. I suspect people are more or less complacent because they don't yet understand or believe the far-reaching implications of these new technologies.

As an Iranian American with extended family still living in Iran, I have witnessed the chilling effect of government censorship and surveillance on individual liberties, but also the power of technology to mobilize people for change.

Thanks to many of the technologies I'll discuss in this book, despite being a chronic migraineur, on most days I am flourishing. I've tried triptans, anti-seizure drugs, antidepressants, brain enhancers, and brain diminishers. I've had neurotoxins injected into my head, my temples, my neck, and my shoulders; undergone electrical stimulation, transcranial direct current stimulation, MRIs, EEGs, fMRIs, and more. On a great day, I'm better than fine—in fact, I'm a little bit ahead of the curve, because the flip side of some of the brain treatments I'll explore is their enhancing effects on other parts of the brain. At times this has given me pause, as I've wondered if my right to treat my migraines is giving me an unfair advantage, or if a stroke of insight is truly my own—or one I should credit to a drug instead.

How technology is used is crucial in defining our cognitive liberty. Exercising a choice to take drugs that changes my brain or to plug in to machines that allow doctors to read it is quite different from if I had been forcibly administered those drugs or if doctors had monitored me without my consent. With our DNA already up for grabs and our smartphones broadcasting our every move, our brains are increasingly the final frontier for privacy.

I worry about our laws' ability to keep up with technological change. Take the First Amendment of the US Constitution, which protects freedom of speech. Does it also protect freedom of thought? Does it give us the freedom to alter our thoughts whenever and however we choose, or can the government or society put limitations on what we do with our own brains? What about the Fifth Amendment? What does it mean to be protected from self-incrimination when the government can hook you up to a machine and find out what's in your mind, whether you want to share it or not? Can companies that we share our brain data with through their applications sell it to third parties? Right now, no laws prevent them from doing so.

Is it so far-fetched to imagine a society in which people are arrested based on their thoughts of committing a crime, as in the dystopian society in the movie *Minority Report*? In Indiana, an eighteen-year-old was charged with attempting to intimidate his school by posting a video of himself shooting people in the hallways . . . except the people were zombies and the video was of him playing an augmented reality video game. Was it really a projection of his intent or, as he insisted, just a game?

Our brains need special protections. If they can be hacked and tracked like all our other online activities and cell phone calls, if our brains are just as subject to data tracking and aggregation as our financial records and online shopping, then we're on the cusp of something profoundly dangerous. But before you panic,

I believe there are solutions—provided that we focus on the right things.

When it comes to privacy protections in general, trying to restrict the flow of information is both impossible and potentially limits the insights we can gain to address the leading causes of disability and human suffering. Instead, we should focus on securing rights and remedies against the *misuse* of information. If people have the right to decide how their brain data is shared, and more importantly, have legal redress if it is misused (say, to discriminate against them in an employment setting or in health care, or education), that will go a long way toward building trust. In many instances, we want to share *more* of our personal information. Aggregated data can tell us so much about our health and our well-being, but we need special protections to ensure our mental privacy.

We must establish the right to cognitive liberty—to protect our freedom of thought and rumination, mental privacy, and self-determination over our brains and mental experiences. This is the bundle of rights that makes up a new right to cognitive liberty, which can and should be recognized as part of the Universal Declaration of Human Rights, which has established mechanisms for the enforcement of international human rights and creates powerful norms that guide corporations and nations on the ethical uses of neurotechnology.

Neurotechnology has unprecedented power to either empower or oppress us. The choice is ours.

PART I

Tracking the Brain

1

The Last Fortress

I hadn't known I would see the future when I agreed to speak at the 2018 Summit of the Wharton Neuroscience Initiative at the University of Pennsylvania. But as soon as Josh Duyan, the chief strategy officer of a company called CTRL-labs, began his presentation, the magnitude of the change and the urgency of the choices we are facing became blindingly clear.

Holding up his hands, Duyan lamented the fact that the extraordinary input capabilities of our brains were tethered to such "limited and clumsy output devices." He noted the step backward we've taken when it comes to typing on our phones, moving from ten fingers to two thumbs. Just imagine how much more efficient we'd be, he said, if we could type with our brains instead or better yet, "operate octopus-like tentacles."

Until that day, I had puzzled over how (and even whether) consumer neurotechnology could go mainstream. The then-existing applications of neurotech that enabled us to play video games, meditate, or improve our focus with our minds seemed like niche applications that were unlikely to motivate people to go about their everyday lives wearing a silly-looking headband.

But the wristband Duyan was describing seemed altogether different. Our brains, he told us, are constantly transmitting signals to our peripheral nervous system—the parts of the nervous system outside of the brain and the spinal cord. CTRL-labs' wristband detects these signals using electromyography (EMG).[1] When I move my hand, for example, the region in my brain called the motor cortex sends an electrical signal to my spinal cord, which distributes a set of signals to the relevant muscles to tell them to move. Where my low motor neurons innervate my muscles, a cascade of activity creates a current (measured in milliamperes) and potential difference or voltage (V, measured in millivolts) that can be detected by the electrodes in the wristband.

With its compact and easy-to-wear form, easy integration into existing wearables—like the smart watches it resembles—and application as an interface to other technologies like virtual reality or swiping a smartphone, this device was different in kind from anything I'd seen. It could offer a significantly better user performance for the tasks currently done by peripheral devices like keyboards and mice.

If people are willing to give up reams of personal data to keep in touch with their friends on Facebook, it seemed likely they would be willing to trade their brain privacy to swipe a screen or type with their minds.

The Last Fortress

The things we think, feel, and mull over in our minds help us define who we are to ourselves and to others. What we choose to share about those things—and perhaps more important, what we choose *not* to share—is fundamental to the intimacy we create with other people.

Until that day in 2018, I believed that our brains were the one place of solace to which we could safely and privately retreat. Your personal diary was always at risk of discovery. If you wrote it on paper, someone could find it and read it; if you typed it on your computer, someone or something could be tracking your keystrokes. But your brain was different. You could think that your friend's new couch is hideous without hurting her feelings. You could think your boss was a clown while nodding affirmatively at his latest pronouncement. You could let your thoughts wander while listening to a boring speaker, fantasizing about your latest romantic interest. Or imagine new ideas or ways of doing things, without having to worry what others would think if your innovations turned out to be duds. You could work through your sexual orientation and decide when and if you would be ready to share that with others. Or you could dare to dream about overthrowing your tyrannical government.

We may soon lose that last realm of privacy. As noted, new technologies collect our brain data to help us become faster, more efficient, safer, healthier, less stressed, and even more spiritual. Just as we exchanged access to our web search history for free and powerful internet browsers, we will have reasons to want to share the brain data these devices collect. To be clear, the data itself is not the same thing as our thoughts and feelings themselves. But powerful machine learning algorithms are getting better and better at translating brain activity into what we are feeling, seeing, imagining, or thinking.

Once we become aware that others can access what we are thinking, feeling, or imagining, we may attempt to censor even our thoughts, lest we be ridiculed or ostracized for having ideas that go against the grain. Worse still, if governments gain the power to track the contents of our brains, they can arrest us and punish us for thought crimes.

I am not alone in my concerns; other scholars are starting to sound the alarm too. The neuroscientist Rafael Yuste has advocated for what he calls neurorights, because, he says, our "brain data may be one of the few remaining bulwarks against fully compromising privacy in modern life."[2] "While the body can easily be subject to domination and control by others," the Swiss bioethicist Marcello Ienca warned, "our minds, along with our thoughts, beliefs and convictions, are to a large extent beyond external constraint. Yet with advances in neural engineering, brain imaging, and pervasive neurotechnology, the mind might no longer be such an unassailable fortress."[3] Dr. Wrye Sententia and the legal theorist Richard Boire have long worried about these issues, having founded the nonprofit Center for Cognitive Liberty and Ethics more than a decade ago.

By now, most people realize that "free" digital services come at the expense of an individual's personal data. While Google originally sought to "bring order to the web" and provide "high quality search results,"[4] it now commands 92 percent of the search engine market in the United States—all the while taking the data we enter into its search engines, web browser, and assets like YouTube and Gmail to create detailed profiles of us that they use to draw conclusions about what different people of different demographics (but increasingly each of us individually) want to see or buy.[5]

Tech companies' business models rest on their ability to sell their understanding of us to others. Google does so through its "real-time bidding" process, which provides advertisers with opportunities to acquire uniquely targeted advertising real estate. Meta does much the same thing, harvesting data on its billions of users and creating psychological profiles of them that advertisers can use to microtarget their pitches.[6] Shoshana Zuboff coined the term "surveillance capitalism" to describe this now ubiquitous phenomenon, characterizing "data about the behaviors of bodies, minds, and things" as "surveillance assets" that can be used for the purpose of "knowing, controlling, and modifying behavior to produce new varieties of commodification, monetization and control."[7]

It's not just tech giants that are commodifying our data, and it's not just advertisers that are interested in it. Consumer data has also enabled a revolution in our understanding of health and disease. The personal genomics company 23andMe, for example, made headlines in 2018 when it announced that it had secured a $300 million deal to share its consumers' genetic data with GlaxoSmithKline.[8] It had already entered into data sharing agreements with other major pharmaceutical companies, including Genentech and Pfizer.

Through its business model, default settings, and privacy policies, 23andMe has included 80 percent of its 10.7 million customers in a database that associates millions of raw genome sequences with consumer demographics and other information, enabling large-scale analyses of genetic diseases and their indicators.[9] That was 23andMe's intention all along, as board member Patrick Chung explained in 2013: "The long game here is not to make money selling [saliva] kits [to collect and report on consumer DNA], although the kits are essential to get the base level data. Once you have the data, [23andMe] does become the Google of personalized health care."[10]

All of which explains why, when I returned home from the Wharton summit, I dived into learning everything I could about CTRL-labs, its products, and the direction it might lead us.

I watched presentations by Thomas Reardon, cofounder of CTRL-labs, describing a future in which our interactions with technology are driven by neural interface. Some of which already exist. Google's *Dinosaur Game* is a feature of its Google Chrome web browser that makes losing internet connectivity a little bit less frustrating. When Google Chrome goes offline and you angrily pound on your space bar, a pixelated *Tyrannosaurus rex* appears. You can use your arrow keys to make the dinosaur run across the side-scrolling landscape and jump over obstacles to earn points. Earn a hundred points, and you'll be rewarded with a screech.

The next time the *T. rex* appears, you could use CTRL-labs' Bluetooth-connected wristband to make the dinosaur jump just by willing it to do so. By maintaining the same mental focus as you work the arrows, the device uses powerful algorithms to translate the electrical activity and brain signals being sent to your hands into signals the computer can understand as commands. "Here's the cool thing," Reardon described. "I don't have to tell you to stop [moving your hand]. What you start to realize is the dinosaur is going to jump whether you push the button or not."[11] Which *is* cool, I thought. But should we really be plugging our brains into Google?

The more I learned, the more certain I became that CTRL-labs would soon be acquired by a major technology firm. It seemed like a natural fit for Apple, as the company could integrate the EMG sensors into its already-popular Apple Watch. In time, the interfaces would allow users to track their sleep, control smart devices in their environment, send text messages just by thinking about it, and even detect signs that they are becoming dangerously drowsy while driving. To my surprise, Facebook's umbrella company, Meta, acquired CTRL-labs instead, in September 2019, paying somewhere between $500 million and $1 billion—one of its most expensive recent acquisitions.[12]

Meta's head of augmented reality (AR) and virtual reality (VR), Andrew "Boz" Bosworth, announced the acquisition on his personal Facebook page. Bosworth explained how the wristbands would become the "universal controller for all your interactions with technology."[13] So far, Meta has showcased typing and swiping with AR and VR as its likely first application, but as Meta founder and CEO Mark Zuckerberg summed it up, "In some ways, the holy grail of all this is a neural interface."[14]

It's easy to see how using neurotechnology as an interface to other technology could fundamentally change our lives. With advances in predictive algorithms, the wristband could anticipate whole words to type from single letters. Reardon calls this "word forming," where "you're not typing. You're kind of forming words in real time and they kind of spill out of your hand. It's giving you . . . choices between words and you quickly learn how to get to the word you want to form." He added, "There would be no difference between how you produce oral speech and how you produce this controlled text flow."[15]

Meta has not achieved word formation at anything close to the rate of speech yet. In demos prior its acquisition, CTRL-labs was only able to achieve forty words per minute (about the same rate as an average typist but significantly slower than our rate of speech, which is about 140 to 160 words per minute). But that was already double the rate achieved by other researchers, and they have undoubtedly made progress in the meantime.[16] With the backing of a company like Meta, real-time neural word decoding is on the horizon.

As for my initial bet that CTRL-lab's EMG sensors would be integrated into the Apple Watch? I was wrong about the acquiring company, but not about its intentions. Meta plans to launch its own smart watch soon. Zuckerberg posted a photo of EssilorLuxottica's chairman, Leonardo Del Vecchio, donning the wristband as part of a joint venture with the smart glasses company. "Leonardo is using a prototype of our neural interface EMG wristband that will eventually let you control your glasses and other devices," Zuckerberg explained.[17] While Meta's first smart watch release may not yet have the EMG sensors, the tech giant promises that future releases will have that integration.[18]

Big Tech Going All In on Brain Decoding

Meta may be leading the big tech pack, but scores of other companies are also in the race to develop neural interfaces. Until now, most have focused on much narrower applications. InteraXon's EEG headset, which I was using to mitigate my migraines, helps consumers meditate more effectively through audible neurofeedback, like the bird's chirping in response to my brain wave activity.

Myontec, Athos, Delsys, and Noraxon offer athletes and sports therapists EMG-generated insights into what's happening with their muscles during training and competitions, such as the rate of force development (a measure of explosive strength), improvements in coordination through training, and symmetry and asymmetry in muscle activation. Control Bionics sells NeuroNode, a wearable EMG device for patients with degenerative neurological disorders like ALS/MND. It enables them to control a computer, tablet, or motor device via the bioelectrical signals that are sent to muscles to trigger movements, even if those movements aren't visible. Kernel offers Flow—a functional near-infrared spectroscopy (fNIRS) device—that looks like a high-tech bike helmet and that measures changes in blood oxygenation levels in the brain to understand and improve its functioning.

But Meta's investment heralds a new frontier for consumer neurotechnology, in which mainstream technology companies use neurotechnology as the new—and potentially primary—way we will interface with all their platforms.

Apple appears poised to make a similar bet, as it has hinted that it will integrate health sensors like EEG into its AirPods, much as it integrated ECG sensors into the Apple Watch.[19] Other neurotech companies are charting the way, making them likely targets for acquisition. Emotiv has launched MN8, earbuds with two-channel integrated EEG sensors.[20] NextSense, backed by Alphabet's moonshot division and spun out as an independent company, believes it has the winning recipe for a brain-health monitoring platform with its EEG-earbuds, and hope to build a "mass-market brain monitor."[21] Apple may be the company's key to doing so. In the spirit of Steve Jobs's Digital Hub, Apple executive Kevin Lynch has extolled the value of multiple devices working together.[22] EEG may be the next device in its wheelhouse.

Snap (the company behind Snapchat) has acquired Paris-based neurotech company NextMind, known for its EEG-based brain wave controller. Snap plans to intergrate the technology into their augmented reality platform, to "monitor neural activity to understand your intent when interacting with a computing interface, allowing you to push virtual button simply by focusing on it," the company explained in their blog post announcing the acquisition.[23]

Microsoft has obtained a patent on an EEG device that allows users to navigate web browsers and apps with their brains and will reward people with cryptocurrency for doing so.[24] Neurable promises "the mind unlocked" with its "smart headphones for smarter focus." And then there are Elon Musk's Neuralink, Thomas Oxley's

Synchron, and Marcus Gerhardt's Blackrock Neurotech (companies we'll look at more closely in chapter 9), which are working on implantable neurotechnology that will be implanted inside our brains. These devices will be far more powerful than any existing wearable neurotechnology—powerful enough to achieve real-time thought and imagery decoding, which is way beyond the capabilities of existing consumer-grade EEG and EMG devices.

But whether worn on our scalps or wrists, or deeply embedded in our brains, all these devices share one striking commonality. Each records our *raw* neural activity—which can be saved, aggregated, and mined for much more than what consumers are using it for. The black box of our brains has been opened. Mark Zuckerberg is right. Neural interface *is* the "holy grail."

Of data tracking by corporations.

"Raw" Brain Data Is Uniquely Sensitive

Suppose you keep a written diary and wish to share a passage from it with a friend. You hand it to them and ask them to read the highlighted passage. Your friend does so and hands it back. Now imagine instead that your friend also made a copy of your diary, which they keep in a file on their desk so they can return to it anytime they want to learn something new about you, whether you intended to share it or not.

Raw brain data captured by EEG, EMG, and other neurotechnology are similar. EEG, for example, records raw brain data—delta (slow waves), theta (medium), alpha (higher), beta (higher still), and gamma (the highest, at 30–80 Hz)—as well as the electrical activity of nearby muscles, electrode motion interference, and ambient noise.

This "raw" data is then fed through software that filters out artifacts and extraneous information, analyzes the brain waves, and picks out the relevant information to return to you. If brain activity is recorded and stored, that same raw brain data can be returned to time and again and mined to learn all kinds of additional insights about you—such as whether you are at risk of stroke or Alzheimer's or ADHD. All without your knowledge.

While you can still choose to use or not use most consumer neurotechnology, once you do use it, you may be revealing far more than you intended: Blinking, the beating of our hearts, sweating—these are all automatic functions that neither require nor follow our conscious wills. More complex automatic brain functions include our visceral or emotional reactions to external events. A scary movie, a passionate kiss, or a painful burn will all evoke automatic reactions that occur outside our cognitive or "rational" thought processing, but that nonetheless leave traces in the brain.[25]

Even emotional states and biases can be decoded. When someone says something that is hurtful, you might choose to conceal your feelings. But your brain still registers them. You might be feeling bored and lonely in your relationship but aren't ready to share that with your spouse. But if your spouse had access to your raw brain data and the tools to interpret it, your brain could give you away.[26] You might work hard to combat your implicit biases while you are at work, but your subconscious still registers them.[27] If you were wearing a consumer headset at the time, your biases could be decoded and made public.

Hackers could even install brain spyware into the apps and devices you are using. A research team led by UC Berkeley computer science professor Dawn Song tried this on gamers who were using neural interface to control a video game. As they played, the researchers inserted subliminal images into the game and probed the players' unconscious brains for reaction to stimuli—like postal addresses, bank details, or human faces. Unbeknownst to the gamers, the researchers were able to steal information from their brains by measuring their unconscious brain responses that signaled recognition to stimuli, including a PIN code for one gamer's credit card and their home address.[28] Neural data could also be intercepted as it is sent to a paired cell phone if it isn't well secured.[29]

"It's happening somewhat faster than we thought," says Howard Chizeck, a professor of electrical engineering at the University of

Washington. Chizeck expects that millions of people will soon be playing online games while wearing brain–computer interface devices. The operator of the game could play Twenty Questions, and measure the automatic brain reactions to what the gamer sees. "I could flash pictures of [gay and straight] couples and see which ones you react to. And going through a logic tree, I could extract your sexual orientation," Chizeck says. "I could show political candidates and begin to understand your political orientation, and then sell that to pollsters." This kind of probing could be accomplished through spyware by a malicious actor but could just as easily be built into popular games and technologies, allowing the manufacturers to surreptitiously collect even more data about us.[30] A recent report claims the Chinese government is already using cutting-edge AI and neurotechnology to analyze facial expressions and brain waves to see if a person is attentive to "thought and political education."[31]

While other personal tracking data has proved strongly predictive of certain things—our purchasing behavior, for example—brain data can reveal our deepest-held feelings or biases, ones we ourselves may not yet have acknowledged. It can even be used to predict how agreeable or neurotic we are by looking at our alpha, beta, and theta bands![32] In the words of the philosopher Sarah Goering and neuroscientist Rafael Yuste, raw brain wave data "could provide access to highly intimate information that is proximal to one's identity," such as our political orientation, sexual orientation, or our tolerance or strategy for dealing with risk.[33]

Many people believe they are shielded from targeted misuse if their information is stripped of identifying information. But our brain patterns may be even more unique than our fingerprints.[34]

Will People Unwittingly Share Their Brain Data?

Will people willingly cede their neural data to corporations? If the world's largest furniture retailer's recent marketing campaign is any indication, the answer is a resounding yes.

Since 2015, IKEA has commissioned contemporary artists to create limited-edition hand-woven rugs, figurines, wall art, and other household objects. Dubbed the IKEA Art Event Collection, the program is intended, in IKEA's words, "to democratize art and make it accessible and affordable for everyone."[35] IKEA was discouraged by the number of buyers who purchased the pieces in bulk so they could resell them through online-auction sites at steeply inflated prices. So, in 2019, it tried something different.

Between May 7 and May 11, 109 rugs were put on display in its Anderlecht, Belgium, store, designed by the likes of SupaKitch, the French street artist; Chiaozza, American play-and-craft sculptors Adam Frezza and Terri Chiao; Noah, a Brooklyn-based multidisciplinary artist; the UK fashion designer Craig Green; Seulgi Lee, a Paris-based Korean contemporary-folklore artist; Virgil Abloh, artistic director of Louis Vuitton's ready-to-wear menswear line; the Japanese-born, LA-based artist Misaki Kawai; and world-renowned Polish pop-culture artist Filip Pagowski. Prospective buyers were told to don an EEG headset so the IKEA (He)art Scanner, a device that was devised in partnership with the advertising and marketing giant Ogilvy Brussels—whose clients coincidentally include Meta—could find out whether they really loved one of the pieces.

"When people looked at each carpet," a representative from Ogilvy explained, "our specially designed algorithm captured and analyzed brain and body reaction data in real time. This data was then promptly translated into an uplift score, which was projected next to the rug they just saw for the first time. If the uplift was sufficient, they could buy the rug. If not, they moved on to try the next one."[36] More disturbing than the blatant intrusion into customers' mental privacy is how blandly unconcerned IKEA customers were about it. Ogilvy claims that not a *single prospective customer* objected to using the EEG device and that "everyone had a great time."

Opening our brains to the outside world, making the information contained therein a target for corporations, governments, and society to use puts all our freedoms at risk. While you may well think, *But, aha! I will just avoid using neurotechnology*, that might not be possible if it's the gateway to goods and services that you already enjoy. And, as we'll explore in the chapters ahead, neurotech may become a requirement in modern workplaces—no wristband, no job.

Though IKEA's campaign is almost certainly a well-intended and harmless gimmick, to me it heralds a chilling future in which society will increasingly encroach upon our brains—to discover anything and everything therein.

Brain Waves as the Newest Status Update

There are, of course, good reasons for us to want to share our raw brain activity data. Along with typing better and interacting with our machines with less friction, it may give us advance warnings for diseases that affect the brain. For that to happen, large sets of brain data must be made available to scientists for analysis. Which means assurances must be put in place so we can share our data without fear that it will be misused.

A few years ago, Kevin Schoeninger and Stephen Skelton learned about EEG studies of monks who meditate and wondered whether consumer neurotechnology could enhance their own practice. Using Muse, InteraXon's EEG device, they took thousands of recordings of their own brains while meditating.[37] Both saw consistent changes in their brain wave activity.

When they launched their ten-week course on meditation, they invited their students to use a Muse headband. But none of their students were able to replicate the changes in brain wave activity that Kevin and Stephen had seen. So, they started experimenting with other devices, among them Flowtime, which records heart rate and breathing alongside EEG. More important, Flowtime allows students to see their raw brain wave activity in real time, and not just the interpretation of that activity that Muse provides, and even the subtle changes in brain waves that a novice could achieve in a short time were visible.

While several students switched to Flowtime, others found a workaround by uploading the Muse data on their brain waves to software from a company called Mind-Monitor.com, which allowed them to view it directly. Kevin told me that both he and his students regularly share their brain wave activity with device manufacturers, third-party software providers, and one another in chat rooms. In fact, thousands of people share and compare their brain wave activity after meditating via social media. On Kevin's prompting, I joined a Facebook group called HeartMind Alchemy Lab, where I was immediately met with screenshot after screenshot of members' EEG activity, which were being commented on and evaluated by other members of the group.[38]

My curiosity piqued, I asked Kevin if he believed that his brain wave activity was any more sensitive than any of the other information he could be sharing on social media. "Um, I haven't really considered that," he replied. I probed a little further and asked if he would have any concerns if the companies who have access to his raw brain wave activity mined it to learn more. "No," he said. "As long as they don't have my social security number and my financial information, I'm not worried. I'm not doing anything that I care if anybody knows." In fact, "everything I'm doing I'm promoting to others!"

Kevin's views align with research results from surveys conducted by my lab at Duke—the Science Law & Policy Lab. Building on a 2014 Pew Research Study on privacy,[39] in 2018 we asked 1,450 US participants to rate the perceived sensitivity of different kinds of information about themselves, ranging from their social security number, health and medication, the contents of their phone conversations, personal relationship histories, and the kinds of information that consumer neurotechnology can reveal—how well they concentrate while working, their emotional states, details about their brain waves, sleep patterns, and alertness throughout the day, the general state of their brain health, and the thoughts in their minds. Just like Kevin, our survey respondents rated their social security number as their single most sensitive piece of personal information. Of the top five items that participants ranked as "very sensitive," only one of our brain variables made the list: the thoughts in their minds. And even those were rated as considerably less sensitive than their social security numbers!

Brain Data Is Being Commodified Too

Whether they are used for meditation, playing games, navigating AR/VR, detecting early seizure onset, tracking tumor development, or detecting the signs of Alzheimer's disease, consumer neurotechnology products all work with apps on a computer or a mobile device. The devices collect raw brain wave and electromyographical signals and filter them through the proprietary algorithms. While the data may have been recorded locally, it ends up on servers operated by companies. And those companies use the data for much more than most of us suspect.

The Flowtime device that Kevin and his students were using? It's manufactured by Hangzhou Enter Electronic Technology (Entertech), a Chinese-based company that has a suite of consumer and enterprise neurotechnology devices with applications for education, psychological health, VR, and the military.[40] Entertech has sold tens of thousands of helmets fitted with EEG sensors to the State Grid Corporation of China—a Chinese state-owned electric utility corporation—so it can measure its workers' fatigue and other brain wave activities in real time on the job. Entertech has accumulated millions of raw EEG data recordings from individuals engaged in various sorts of activities, from mind-controlled video game racing to working and sleeping.[41]

Entertech collects much more than raw brain wave data from its users. It records their personal information when they create their accounts—such as their birthdate, gender, height, weight, and in some cases, mobile telephone numbers. When location services are enabled on the app, Entertech collects GPS signals, device sensors, WiFi access points, and cell tower IDs. It tracks information about other devices, computers, and services a person may be using, including their IP addresses, browser types, languages, operating systems, referring web page, other pages visited, and cookies they may have installed. If a user connects their services to Meta or Google, Entertech will collect their email addresses, and those of their friends. And it associates all this data with a user's raw brain wave activity.

If you read Entertech's privacy policy, you'll see that the company plainly discloses its intention to use all this information and to share aggregated or de-identified data, including brain wave activity and its interpretation of that data, with third-party partners.[42] Given all of that, it's not surprising that in November 2018, SingularityNET announced that it had entered a partnership with Entertech to analyze data gathered from Entertech's EEG measurement products using its AI platform. While it's unclear what they ultimately hope to learn from the data, they have publicly stated that they aim to "allow people to regulate their states of mind better so as to guide their daily lives and their meditative experiences," and to "help enterprise employees better monitor and self-regulate their states of consciousness on the job to improve both their job performance and their satisfaction."[43] As we'll see in the chapters that follow, employers are already making use of their insights. And so is the Chinese military.

Other consumer neurotech companies are similarly open about their commodification of the raw brain data they collect. Developed in 2011 by students at the MIT Media Lab, Multimer's MindRider EEG bike helmet was originally designed to read a bicyclists' brain waves so it could alert motorists and the cyclists to their mental states with colored lights—such as whether they were focused or drowsy, or nervous because you were driving too close.[44] The company's founder, Arlene Ducao, then went on to create Multimer Data, which received its first round of seed funding in 2016, to mine the data MindRider collects to create maps of cities with location-specific insights gathered from users' brains—like the ideal site for a retail establishment, or where to place a billboard so that it has the least visual distractions surrounding it.[45]

Some companies have progressive privacy policies. Mind-monitor .com promises not to "collect your personal brainwave data," and says that the "brainwaves recorded by Mind Monitor are stored only on your personal device and [we] have no access to them."[46] But its approach is a rare exception. Most corporations claim unfettered rights to our raw brain-activity data, and to record, store, and mine it.

As for Meta? Commoditizing consumer data *is* its business model. While it emphasizes that CTRL-labs' EMG device collects information locally, at muscle junctures rather than the brain, EMG data is no less sensitive than raw brain data.

Whether you're washing your hands, developing an imperceptible tremor as you progress into the earliest stages of a neurodegenerative disease, are forming words you decide not to type, or are engaged in intimate activity using your hands in your bedroom, your EMG wristband can recognize it. Which means Meta can too.

That's why it's high time that consumers wake up and start to learn about the unique sensitivity of their brain-activity data—and why our mental privacy is worth defending.

Brain Data Varies in Sensitivity, So Governing It Should Too

In the summer of 2020, Chicago police investigated a homicide recorded on a home security camera. The recording shows two men fighting in front of a Rogers Park apartment building. One punches the other in the head. The punched man staggers, then pulls a gun and shoots his assailant. The gunman's face is never visible, but the video captures his voice.

Based on an anonymous tip, the police identify a suspect, but even the voiceprint can't connect him to the crime.[47] Could neurotechnology help?

I started to grapple with the possibility of our brains being used as evidence against us in a criminal trial in a pair of law review articles.[48] I laid out a spectrum of neurological evidence that could be used—from personally identifying to automatic, memorialized, or silently uttered information gleaned from our brains. In doing so, I realized that not all brain evidence is equally sensitive, and perhaps not all of it should be off-limits.

We might begin by obtaining physically identifying evidence from the suspect. The police knew that the gunman suffered a blow to the head, so brain activity could reveal head trauma and how recently it had occurred, even when a visual inspection of his head might not. Much like a blood sample, a saliva swab of DNA, or a dental exam, brain activity of this sort can provide identifying evidence about a suspect's physical state without revealing information about his mental processes.

But the police would need more than identifying information to prove the gunman was involved. They might, for example, try to measure his visceral or emotional reactions to photographs of the victim. This kind of automatically generated unconscious brain information would be more sensitive than traces of a trauma, but still less revealing than the most intimate information we might obtain.

Our brains, after all, preserve not just our present thoughts and mental experiences but also our memories of everyday encounters. These include the people we have met: the timbre of their voices, visual imagery we encountered, and episodic memories of our experiences. The police could play the tape of the victim's voice and see whether the suspect's brain reveals recognition.

The most intimate information we might decode from the suspect's brain, which would most directly rob us of our control over what we share with others, would be the silent utterances, thoughts, visual images, or statements present in his mind. The respondents to my labs' survey of a nationally representative sample of US residents ranked the thoughts and images in their mind as highly sensitive information, implicitly recognizing the difference in sensitivity between silent utterances and memories in our minds versus the automatic functional information and identifying characteristics of the brain that they ranked as less sensitive.

As we endeavor to define the contours of mental privacy included within the right to cognitive liberty, we should recognize that not all neural data is equally sensitive. While the right to mental privacy should protect the entire spectrum of neural data—identifying, automatic, memorialized, and silent utterances—the individual interest in mental privacy should be the most unyielding when silently uttered or memorialized information is sought. When automatic brain functioning or identifying information is sought instead, mental privacy may at times yield to societal interests.

As vigilant as we should be when it comes to mental privacy, there are reasons we shouldn't make our brains completely off-limits, and instead balance the interests at stake. We don't want to unnecessarily constrain neuroscience research by prohibiting the use of any neural data merely to ensure the most sensitive brain data is protected.

There Is Value in Sharing

Balancing our individual interests against society's will help us define our rights. We'll explore this process in the context of the modern workplace, education, and other settings in the chapters to come. But individually and collectively, we stand to benefit a lot from what we can learn from our brain wave activity. Large data sets of brain wave activity could enable the promises of precision medicine—from discovering the causes of neurological disease to individualized treatments for afflictions of mind, helping us to eliminate some of the leading causes of human suffering today.

Data sets like those collected by Entertech are unique in that they capture neural data of individuals "in the wild" rather than in more limited and controlled research settings. While such data has limitations in quality because it isn't well controlled (environment and other factors aren't as consistent as they would be in a laboratory), and people moving about their everyday lives with neural interface devices may give rise to many recorded artifacts from movement, this kind of data nevertheless offers unprecedented opportunities for understanding brain functioning across individuals and in the real world. And that's precisely *why* we need large data sets—so we can understand how noise, device variability, and user differences impact the accuracy of the measurements, and improve the devices and the algorithms that they use.

There are very few big data sets that researchers can use to study the neural activity of healthy individuals in their everyday lives. There have been some large-scale efforts to promote brain data sharing through organizations like the International Neuroinformatic Coordinating Facility in Stockholm and the Neuroimaging Informatics Tools and Resources Collaboratory in the United States.[49] But when it comes to raw EEG, EMG, and fNIRS data, "the vast majority of human neuroscientific studies use a very small number of participants employed in very specific tasks," as Dr. Jonathan Touryan, an army scientist for the US Army Combat Capabilities Development Command's Army Research Laboratory, puts it. Touryan's laboratory has teamed up with the University of Texas at San Antonio and Intheon Labs to develop a first-of-its-kind mega-analysis of EEG data.[50]

EEG is decoded using powerful machine learning approaches, including what's known as neural networks and support vector machines.[51] Widespread use of consumer neurotechnology will allow for the creation of bigger and better data sets that can train more advanced machine learning algorithms. With larger and more diverse data sets, scientists can identify universal features that link neural activity to different brain states and tasks. This isn't easy to do in EEG and EMG, given all the different devices that are in use—with different configurations, numbers, and locations of electrodes—the different tasks people are performing while wearing them, and the different ways those tasks are annotated by the software powering them.[52] But with mega-analysis, it may be possible. This would allow for improvements in the accuracy and predictive quality of consumer neurotechnology, bringing significant benefit to the individuals who use it.

But we can realize these benefits only if our brain data isn't coopted by a few big tech giants like Meta, which may choose to keep it proprietary, for its own purposes. To prevent that data grab from happening—and to protect individuals from the misuse of their personal information—we need to establish new norms, put into place clear rules and regulations to protect our mental privacy, and see that they are strictly enforced.

Most people will be reticent to share their neural data unless they are assured that it won't be used against them. As we explore the growing use of neurotechnology in employment, education, and the government in the chapters ahead, we'll look at the additional context-specific protections required to enable people to share their neural data without facing discriminatory consequences.

Sharing our neural data creates unprecedented possibilities to solve some of the most challenging health issues we face. But it also introduces unprecedented risks to our right to our own vulnerability, which is one of the many reasons we need to explicitly define the right to mental privacy. In doing so, we can specify precisely who can track our raw brain activity, when they can do so, and for what purposes.

The Right to Mental Privacy

It isn't too late to secure individuals a right to mental privacy. By recognizing a new right to cognitive liberty and defining its contours, we can enjoy the benefits of neurotechnology while preserving a space for mental reprieve.

Securing mental privacy to individuals will require us to update and broaden our contemporary understanding of the bundle of rights included within cognitive liberty in international human rights law—including mental privacy, freedom of thought, and extending the collective right of self-determination to individuals. In later chapters, we'll explore self-determination and the absolute right to freedom of thought (which overlaps with mental privacy to protect *any* unauthorized access to the silent utterances and memories in our brains). Mental privacy protects brain data, too, but less stringently than the right to freedom of thought recognized in international law, since it is a relative right, meaning that it will yield to societal interests when warranted.

The United Nations (UN) has been credited for launching the international human rights movement, galvanized by the atrocities committed during World War II. Although the normative foundation for human rights began well before the founding of the UN in 1945, the adoption of the Universal Declaration of Human Rights (UDHR) in 1948 was a crucial milestone. The declaration outlines thirty rights and freedoms necessary to secure human dignity to every person and forms the basis for modern international human rights law.

The UDHR established important global norms about human rights and created moral obligations for governments, corporations, and individuals to follow and international condemnation when they fail to do so.[53] The UDHR is also the source of international treaties—including the International Covenant on Civil and Political Rights (ICCPR)—that define freedom of thought and other human rights with greater specificity. These treaties create international courts and monitoring bodies to review whether governments are respecting these rights. The UDHR and human rights treaties are also incorporated into national laws around the world to help make the rights enforceable.[54]

The Swiss bioethicist Marcello Ienca and the law professor Roberto Adorno have called on the UN to recognize a new human right to mental privacy "to protect any bit or set of brain information about an individual recorded by a neurodevice and shared across the digital ecosystem." They see this as a *relative* right that "can be limited by certain circumstances, provided that some restrictions are a necessary and proportionate way of achieving a legitimate purpose."[55] A group of ethicists and legal scholars called the Morningside Group have similarly called for a new right to mental privacy, as an absolute right, where "any interference with it by States without consent should be considered *de facto* cruel, inhuman, or degrading treatment."[56]

A new right to mental privacy may not be necessary, however. Both privacy and freedom of thought are *already* protected by the UDHR. Article 12 of the UDHR states that "No one shall be subjected to arbitrary interference with his privacy, family, home or correspondence, nor to attacks upon his honour and reputation." And Article 18 affirms that "Everyone has the right to freedom of thought . . ." While mental privacy is not explicitly mentioned in these articles, it is widely recorgnized that the interpretation of human rights law evolves as we discover and deliberate about novel challenges to human dignity.[57] As changes in society and technology reveal gaps in our interpretation of existing human rights, we can and should update application of human rights in to these new contexts.

Standard setting in international human rights law is an important first step to establishing social norms and moral obligations for governments, corporations, and individuals to follow. Norms have a powerful effect in their own right—actors who violate human rights norms often face international scrutiny and accountability for doing so. But merely updating those rights to make explicit that they include mental privacy will not transform behavior. For standards to become effective, there must also be pathways for enforcement within countries.[58]

In other words, *how* the right to mental privacy is implemented will become just as important as its recognition. Implementation will require corporations, governments, academics, and the public to join in deliberating and defining the requirements for the responsible use of neurotechnology and neural data in society.

For example, when Joseph Cannataci, the special rapporteur on the Right to Privacy, presented the final report of his six-year mandate to the UN Human Rights Council in 2021, he advocated for implementing norms for data management to protect society as the use of artificial intelligence proliferates. He also urged minimizing data use to the original purpose for which it was collected, securing people against the use of data to discriminate against people, and improving data accuracy.[59]

Cannataci's report and proposals are a good starting point, but they don't go far enough. This book offers a comprehensive blueprint to recognize a new international human right to cognitive liberty and the right to mental privacy it includes. The chapters ahead discuss how the specific contours of mental privacy will apply in different contexts, and the need to adopt data management practices within countries for the protection of neural data. At a minimum, we should require that corporations be transparent about the neural data they are collecting. Apple has paved the way by developing "data nutrition labels" for the apps in its store, arming users with knowledge of not just how an app is used but how it's using *them*.[60] We need similar nutrition labels on neurotechnology devices and their related apps so we can know what kinds of neural data are being collected, and how that data is being *used*. Recognizing a human right to mental privacy in international law will create the powerful norms to incentivize more corporate actions like this one. But greater transparency is not enough.

We must also require through laws, regulations, and norm enforcement that corporations limit further processing of raw neural data to prevent them from mining it for "sensitive" data—including memorialized and uttered information, and even automatic functions of the brain that are more closely aligned with our sense of self. Any access to brain data will violate mental privacy, so it should occur only when consumers have explicitly opted *in* to that information being processed and only if there is a compelling justification for its use. While opt-in creates much greater friction for companies, it is not only justified but also essential to secure our mental privacy.

Corporations must also create individual user-based controls on the devices and applications themselves. These controls should give users full transparency about what is being collected, stored, and shared, and allow them to switch the devices on and off so they can control what data is being collected and when. EMG watches and EEG earbuds, for example, should have Power-On and Power-Off switches that will allow users to wear them continuously without having to worry about what neural activity is being collected. Similarly, apps should give users the ability to store their raw neural data for processing locally, and to have that data overwritten continuously rather than stored indefinitely on the companies' servers for further processing by themselves and their partners.

Individuals should also be empowered to share their raw neural data in a de-identified and aggregated form. This will require society to implement standards against the discriminatory use of this data, and to ensure that individuals have rights of redress if that data is misused.

Developing a human right to mental privacy and implementing it across the world will require us to define the kinds of analyses of raw neural data that are permissible through a continuing process of democratic deliberation, a truly inclusive one, that allows for town hall–style forums to discuss issues like data mining. This process will increase public awareness and create a platform for sharing ideas and values. And it will empower participants to work in partnership with decision-makers (including corporations) to implement specific policies that will secure mental privacy to individuals in the domains they care about the most.

Mental privacy is a critical aspect of cognitive liberty. But like all privacy interests, it is not absolute. People can and should have the right to give others access to their brain activity. There will be times when we will want to do so, to promote research or in exchanges for goods and services. And there will be times that society will demand tracking of our brain activity when the lives of others are at risk.

2

Your Brain at Work

Kerry Cuthbert was suspended from his job while investigated by his employer after he made a 1,470-mile run, all the way from Wisconsin to Florida, in one twenty-hour shot.[1] That well exceeded the maximum number of consecutive hours a long-haul trucker is allowed to drive to prevent injury to himself or others. Kerry's employer, International Logistics Group, only learned about his risky venture when he later bragged about it on social media.[2]

For the past decade, SmartCap Technologies has sold technology to companies that allows them to monitor their employees' brains. From Australia to the Americas and parts of Africa, more than five thousand companies worldwide in mining, construction, trucking, aviation, railway, and other industries use SmartCap to ensure that their employees are wide awake.[3]

Powered by the LifeBand, a headband with embedded EEG sensors that can be worn alone or integrated into headwear like a hard hat or cap, brain wave data is processed through SmartCap's LifeApp software, which uses proprietary algorithms to provide a risk assessment score based on the wearer's ability to resist sleep. LifeApp presents fatigue levels on a scale from 1 to 5, from hyperalert to involuntary sleep. When the system detects the wearer is becoming dangerously drowsy, it sends an early warning to both employee and manager.[4]

My initial instinct was to be troubled by the idea of an employer tracking my brain activity in this way. But the more I've thought about it, the more I've come to believe the issues are far more nuanced. To be sure, the technology raises significant privacy concerns for employees. It's even more troubling when they don't know what brain data is being collected or how their employer will use it. Whatever is gained in workplace safety or productivity could be offset by the potential loss of trust, engagement, creativity, and empowerment. And yet, I think it's possible that at least for some applications, we should embrace brain monitoring at work. It's hard to argue that an employee's abstract mental privacy trumps society's very real interest in knowing that the driver of a forty-ton truck barreling down the highway is wide awake and alert.

We still have time to get this right, to thoughtfully balance the interests of employees, employers, and society. But to do so, we need to develop a new set of human rights norms rooted in cognitive liberty.

The Rise of Workplace Surveillance

By now you understand how corporations are using our data to predict our preferences and desires as consumers. But you may not realize that they are also using technology to analyze our behaviors as employees, tracking our movements and keystrokes, and now even our brains.[5]

The UK grocery chain Tesco was one of the first companies to employ such surveillance. In 2013, it required workers in its grocery warehousing facilities in Ireland and the UK to wear armbands that tracked their productivity. Every time a worker picked up a piece of inventory and moved it from one place to another, the armband logged those movements. It also tracked bathroom breaks.

The pushback was loud and angry. Tesco workers felt like Big Brother was watching them, and they *really* didn't like it, even though it made their lives easier in some ways, like not having to carry pens and papers or log inventory manually. But as one former employee revealed, workers lived in constant fear of being called in by management for low performance scores. Even the "the guys who made the scores were sweating buckets and throwing stuff around the place," he said.[6]

Maybe you're thinking, *Too bad! Employees shouldn't take unscheduled breaks!* Or that if you were in their shoes, you would just quit and go elsewhere. But most of those low-wage workers didn't have the luxury of job mobility. And it didn't take long before there was nowhere else to go, as surveillance became more and more common at workplaces worldwide.

That's what keeps Stacy Mitchell, codirector of the Institute for Local Self-Reliance, an organization that focuses on the risks of corporate power, up at night. Stacy told me that she believes "technology has become a tool for powerful companies to exert even more power over their workforce." Which means that the average worker has "fewer options and less leverage" while corporations have an "incredible ability to control and intimidate" workers through surveillance tools.[7]

Much of Stacy's research has focused on the growing abuses of surveillance at Amazon's global distribution centers. Amazon also has a patented wristband to track its employees,[8] as well as a broad surveillance program called AWS Panorama, which uses machine learning algorithms to process images from cameras, including the security cameras that are ubiquitous at most modern worksites. "Are people walking in spaces where they shouldn't be? Is there an oil spill? Are they not wearing hard hats? These are real-world problems," a senior Amazon executive said while describing the myriad ways that AWS Panorama can be used.[9]

The result, Stacy explained, is that workers feel "constantly rushed and prodded and prodded. And the rate at which they have to do tasks is forever increasing. As soon as they think they've mastered the existing rate, it goes up higher and higher." Which means that companies like Amazon are not only "controlling people moment to moment, but also creating an environment where they can never actually succeed."[10] The mounting pressure led Chris Smalls to help Amazon workers in Staten Island to successfully unionize, and countless workers at other companies—from Walmart to Target to Dollar General—have reached out to him for help in doing the same.

Surveillance goes well beyond warehouses and extends to knowledge workers, too—a trend that began well before the COVID-19 pandemic but has grown exponentially since. In a recent survey of employers with remote or hybrid workplaces and their employees, 78 percent of employers admitted to using monitoring technology to track their white-collar employees.[11] Using programs that are pejoratively called "bossware," they take regular screenshots of employees' work, monitor their keystrokes and web usage, photograph them at their desks, and track their social media usage, all in the name of increased productivity.[12]

Now employers are starting to look directly into employees' brains for signs of fatigue and wandering attention. The data they are collecting will undoubtedly reveal rising stress levels in their workforce.

Tracking Fatigue at Work

In 2019, Tim Ekert, the CEO of SmartCap Technologies, boldly proclaimed that his company's headgear would "transform the American trucking industry" by reducing fatigue risks for drivers and employers.[13] Why was he so confident? Because SmartCap had just concluded a successful three-month partnership with a US-based trucking company that yielded some very impressive results. Eighty-seven drivers at two terminals in the company's refrigeration division had worn SmartCap devices to monitor their fatigue levels while working.

One terminal operated primarily at nighttime, while the other operated 24/7. The participating drivers logged over 8,730 hours and 520,000 miles without any fatigue-related accidents while realizing a 62.4 percent reduction in fatigue-related alarms.[14] That's a remarkable achievement for what seems like a pretty minor ask.

This was far from SmartCap's first rodeo. In Australia, where SmartCap is headquartered, mining, aviation, and gas companies regularly use its devices.[15] At the Hunter Valley Operations in New South Wales, SmartCap is mandatory for truck drivers.[16] BAM Nuttall, the British subsidiary of the Dutch Royal BAM Group, which employs well over twenty-five thousand people in its civil engineering and construction divisions, equips many of its employees with SmartCap gear.[17] A South American mining company replaced its camera-based monitoring system with SmartCap's EEG headsets.[18] A large copper-mining operation in East Asia requires all its drivers of heavy vehicles to wear SmartCaps.[19] All in all, upward of five thousand companies in mining, construction, trucking, aviation, railway, and other industries worldwide use SmartCap.

SmartCap is so successful because of its promise to reduce the catastrophic costs that drowsy employees levy on businesses and society. In Chicago, a speeding train jumped its tracks at O'Hare International Airport after its driver fell asleep and careered onto an escalator, injuring thirty-two. The driver admitted it wasn't the first time she had dozed off at the controls.[20] In New York, a sleep-deprived train engineer nodded off while operating a Hudson Line commuter train from Poughkeepsie to Grand Central Terminal in New York. The train took a 30 mph curve at 82 mph and derailed, killing four people and injuring seventy, and causing millions of dollars of damage.[21] In Citra, Florida, an engineer fell asleep while operating a train that was pulling a hundred carloads of phosphate and crashed head-on into a coal train. Thirty-two cars left the tracks, spilling 1,346 tons of coal, 1,150 tons of phosphate, 7,400 gallons of diesel fuel, and 77 gallons of battery acid.[22] Although aviation accidents are much less common, at least sixteen major plane crashes have been blamed on pilot fatigue in the past few decades.[23]

Until recently, the state of the art for monitoring driver fatigue was telematics technology—hardware and software built into vehicles that inferred the driver's drowsiness from the way they drove. Take Mercedes-Benz's smart-sensor system. It "reads" the driver during their first few minutes behind the wheel, analyzing their steering movements and speed under ordinary conditions. If the driver later makes a number of minor steering errors that are quickly and repeatedly corrected, the sensors, computer algorithms, and software program evaluate internal and external vehicle conditions to determine if the driver is drowsy. If the answer appears to be yes, the vehicle issues an alert to the driver to take a "brake."[24] Other manufacturers offer similar systems.[25] While they represented a remarkable advance when they were introduced, their accuracy rates are only between 63 and 75 percent.[26]

Newer telematic systems use in-vehicle cameras. These systems can be highly accurate but come at a high cost to individual privacy. One system, for example, uses facial recognition technology to analyze whether the driver is looking ahead at the road with her eyes open.[27] Others use infrared cameras to monitor the driver's head and body positions, as well as facial expressions and hand gestures.[28] The German technology firm Bosch Global created an interior monitoring system that uses artificial intelligence to determine when a driver is distracted, when their eyes are becoming heavy, and even when they are looking at the passengers next to them or in the rear seat.[29]

EEG systems like the one used by SmartCap can be implemented in ways that are much less intrusive, as they are focused solely on brain wave activity. As our alertness wanes, our brain waves show a rise in theta and alpha activities and a fall in beta activity.[30] Alpha brain wave activity alone can be used as an indicator of impending fatigue.[31] Despite some important limitations—like the small number of electrodes in consumer EEG devices, which reduces the overall quality of the data being recorded; the computational complexity of classifying brain wave signals as "drowsy"; and the "noisiness" of the signal that is being captured[32]—even the worst-performing EEG devices tend to detect fatigue earlier and more reliably than traditional telematic technology, and the best performing ones do as well as, if not better than, facial recognition and in-vehicle cameras.[33]

EEG systems can be used in other employment settings where fatigue impacts safety—from factory floors to air traffic control towers, operating rooms, and laboratories. And perhaps they should! Fatigue reduces our motivation, concentration, and coordination; lengthens our reaction times; undermines our judgment; and impairs our physical and mental abilities to carry out even the simplest tasks. The result is more than $136 billion per year in productivity losses from corporate bottom lines.[34]

As neurotechnology and the algorithms decoding them continue to improve, EEG-based systems will become the gold standard in workplace fatigue monitoring. Not just employers but society as a whole may soon decide that the gains in safety and productivity are well worth the costs in employee privacy. But how much we ultimately gain from workplace brain wearables depends largely on how employers leverage the technology. Will employees receive real-time feedback from the devices so they can act on it themselves? Will managers directly monitor employees' incidence of fatigue? If so, will they use that information to improve workplace conditions? Or will it justify disciplinary actions, pay cuts, and terminations of employees who suffer from fatigue more often? The answers to these questions will shape the future of brain-fatigue monitoring.

Given the lack of societal norms and laws regarding brain monitoring, for now each company creates its own.[35] Some use SmartCap and similar technologies to optimize employees' working conditions; others may use them punitively, which is consistent with how employers generally use workplace surveillance. In one recent study, 26 percent of employers had fired employees for internet misuse, 25 percent had terminated employees for email misuse, and 6 percent had fired employees for personal use of office telephones.[36]

While neurotech companies tout their technologies as productivity enhancers, their implementation runs the risk of backfiring. Creating clear and generally accepted norms for the use of employee surveillance is crucial to its acceptance in the workplace. One in three people already mistrust their employers, and employee trust is an essential ingredient of corporate success. Employees in high-trust organizations are more productive, have more energy, collaborate better, and are more loyal than those who work for employers they don't trust. Employees in low-trust companies feel disempowered and become disengaged.[37] Corporations in the United States are estimated to lose between $450 and $550 billion each year because of employee disengagement.[38]

Watching Our Wandering Brains

I used to be able to undertake marathon writing projects. I'd put on my most comfortable clothes, make myself a huge vat of hummus for breakfast, lunch, and dinner, and dive in to researching and writing my newest piece. A few weeks "into the zone," and I would come up for air with a draft in hand. That all changed when I had children. Constant interruptions became my new normal. I soon realized that I needed to fundamentally change the way I work to regain my productivity.

After some fits and starts, I discovered a new method that better enabled me: the Pomodoro technique, invented by Francesco Cirillo in the 1980s. You choose a task to focus on, set a timer for twenty-five minutes (the original timer was shaped like a tomato, or *pomodoro* in Italian), and focus on one and only one task. Then you take a five-minute break and begin again.[39] My children learned to wait for the timer before they interrupted me, and I learned to write in twenty-five-minute bursts.

We don't all have the luxury or ability to focus for long stretches at a time, which is something that Olivier Oullier, the president of the bioinformatic company Emotiv, based in San Francisco, California, thinks neurotechnology can address. At the outset of a recent talk he gave at the Fortune Global Tech Forum, Oullier acknowledged that we are "not equal when it comes to focusing. Some people can focus very, very deeply for forty-five minutes. Others for two hours."[40] But the solution he offered was a far cry from my desktop tomato.

Oullier was unveiling the MN8, Emotiv's enterprise solution for attention management. The MN8 looks like standard earbuds (and can in fact be used to listen to music or participate in conference calls). But with just two electrodes, one in each ear, it allows employers to monitor employees' emotional and cognitive functions in real time.[41]

Emotiv teamed up with the German software company SAP SE to create Focus UX, a system that reads employees' cognitive states in real time and shares personalized feedback on their attention and stress levels with both them and their managers. SAP claims this will create a more responsive workplace environment in which computers help employees focus on what they are best "able to handle at that moment."[42]

Oullier described a hypothetical data scientist who is wearing the MN8 earbuds. She's spent several hours videoconferencing with her team and is now reviewing codes. Alpha brain wave activity has been used to index the attentive state in her brain. Higher alpha power is correlated to mind wandering or inattention, while lower alpha power has been correlated with a more attentive state.[43] The proprietary algorithm sees that her attention is flagging, so it sends a message to her laptop: CHRISTINA, IT'S TIME FOR A BREAK. DO YOU WANT TO TAKE A SHORT WALK OR DO A FIVE-MINUTE GUIDED MEDITATION TO RESET YOUR FOCUS? [44] Employers can use the data to evaluate individual users' cognitive loads, compare them across their workforce, and make decisions about how to optimize their workforce for greater productivity throughout the day. And of course, to make promotion, retention, and firing decisions.

Other companies offer similar technology, such as Lockhead's CogC2 (for Cognitive Command and Control), which provides companies with real-time neurophysiological workload assessments so they can "understand the performance cycles of individuals and teams" and "optimize their workforce for increased productivity and improved employee satisfaction."[45]

All this might remind you, as it did me, of the incredibly unpopular "attendee attention tracking" feature that Zoom rolled out at the height of the global pandemic, which informed meeting hosts when an attendee minimized their Zoom window for more than thirty seconds (maybe it's more like that feature on steroids). Public outrage and negative media coverage quickly led Zoom to remove the feature, which was a relief to me as I multitasked through my Zoom meetings or turned away momentarily to attend to one of my children.[46]

In case you are thinking, *Well, even if my employer tries to measure my attention, they won't know what I am paying attention to,* think again. Research on workplace engagement funded by the Bavarian State Ministry for Education and Culture found that with EEG, it is now possible to classify the *type* of activity an individual is engaged in—central tasks (e.g., programming, database, web development), peripheral tasks (e.g., setting up a development environment, writing documentation), and meta tasks (e.g., social media browsing, reading news sites).[47] As pattern classification of brain wave data becomes ever more sophisticated, employers will be able to tell not just whether you are alert or your mind is wandering but also whether you are surfing social media or developing code.

Your employer might even nudge you back to work when your mind starts to wander. The MIT Media Lab developed a system called AttentivU, which measures a person's engagement in real-time via an EEG headband. A scarf provides subtle haptic feedback in the form of vibrations whenever the wearer's engagement declines. Researchers found that people who received haptic feedback logged higher EEG alertness scores than those who didn't. While the Media Lab Group were excited about their results, they acknowledged the risk for misuse, saying they hoped that "no one will be forced to use this system, whether in work or school settings."[48] Given our current workplace trends, that hope seems unlikely to be realized.

But it doesn't have to be that way. If some employees choose to use neurotechnology voluntarily to improve their productivity, others may clamor for it too. Empowering people with tools to improve their productivity while giving them control over the data they generate could allow them to reap the benefits of better time management without any sacrifice of their autonomy. Just like my pomodoro timer! As with other neurofeedback, self-monitoring for productivity can help us establish more positive work habits as we learn when and why our attention wanders.[49] This can be empowering in other ways too. The more productive we are, the greater our bargaining power to demand higher wages and better working conditions.[50]

If brain productivity technology is imposed on workers by management and used punitively instead, we should expect employees to object to its use. Some are already doing so. Unionized workers at Rio Tinto's Hail Creek mine in Queensland, Australia, refused to wear SmartCaps because of just those concerns in 2015.[51]

Those concerns resonate with my own research findings. My lab conducted a follow-up survey to our study on the sensitivity of brain data that I discussed in chapter 1. We asked 110 Americans what concerned them the most about others having access to their brain data. Far and away, their number one fear was the uses their employers might put it to, much more than governments, insurance providers, law enforcement, advertisers, hackers, or family and friends.

They have good reason to worry! Existing laws do little to protect employees from workplace surveillance. The European General Data Protection Regulation (GDPR), which applies to EU residents, creates a floor that many countries now abide by. But the GDPR and applicable local privacy legislation only requires employers to have legitimate reasons for collecting employee data. Those can include public health and workplace safety, but they also extend to workplace efficiency and productivity.[52] The result is that employers who decide to adopt workplace brain monitoring are generally free to do so.

Data-protection statutes across the world strongly favor freedom of contract between employers and employees and limit government oversight that would interfere with the terms and conditions to which employers and employees agree. Take Article 329 of the Constitution of Ecuador, which forbids the use of discriminatory instruments that affect people's privacy.[53] At first glance, it appears reasonable, but it says nothing about the use of technology in the workplace for non-discriminatory uses. In most countries, such as Mexico and South Korea, if an employee consents to the use of surveillance technology, the requirements of workplace privacy are satisfied. In practice, that means so long as brain monitoring is fully disclosed in employee handbooks and privacy notices, an agreement to work is tantamount to consent. Even Article 19 of the Chilean constitution, which secures to individuals a right to physical and mental integrity, can be abridged with consent by an employee.[54]

And yet, I suspect that many employers will discover that brain monitoring won't be worth the squeeze on their employees. Monitoring brains for focus may increase stress and undermine employee morale, which could make employees less rather than more productive.[55] Employees have been known to engage in counterproductive behaviors when they dislike surveillance technology, such as deliberately manipulating surveillance systems, taking great pains not to be monitored, or even falsifying their work product to meet performance requirements.[56] "They basically can see everything you do, and it's all to their benefit," said Courtenay Brown, who works in a giant refrigerated section of the Avenel, New Jersey, Amazon Fresh warehouse. "They don't value you as a human being. It's demeaning." The growing belief by workers that surveillance creates inhumane working conditions is fueling unionization efforts across the country.[57]

I witnessed these effects first-hand recently. Like many others, I have come to rely heavily on the gig-economy workers who deliver groceries and other essentials to our home. It's not uncommon for our family to place several grocery orders per week with Instacart, a company that operates a grocery delivery and pickup service in the United States and Canada, from different venues in tow. One day, I placed a same-day delivery order only to realize that I had left off a few essentials. Since I had already missed the time window to update my shopping cart, I placed a second order for the next two-hour window. Twenty minutes later, I received a panicked text message from an Instacart shopper, Shannon, saying she had been working on my first order when she was assigned the second one. The problem, she explained, was that if she delivered both my orders together in the later time window, she would be penalized for delivering the first order late. "Could you call in and change the delivery time for the first order?" she asked.

Instacart employees receive weekly productivity scores based on factors like the number of on-time and late deliveries they make, how long it takes them to pick each item in the store, and the percentage of customers they correspond with. Instacart's laser-like focus on productivity has improved its profit margin, but its shoppers live in constant fear of being fired over situations that are often out of their control, like the quandary I had inadvertently put Shannon in.[58] To meet the time constraints constant monitoring has introduced, some workers skip bathroom breaks, relieving themselves in bottles and plastic bags.[59]

When our attention also becomes the currency of productivity measurements, employees may be driven to similarly extreme measures, such as attempting to combat even brief periods of mental downtime. This can substantially erode the quality of their contributions. Research across nine hundred Boston Consulting Group teams in thirty different countries found that downtime is critical to employee success, as it increases alertness, improves creativity, and leads to greater output quality.[60] Creative ideas and solutions depend as much on minds wandering as staying on task. Albert Einstein and Isaac Newton both famously claimed that their task-unrelated thoughts were critical to the biggest problems they addressed in their work.[61]

Decades of medical and psychological research show the negative impact of high job demands and low job control on workers' health. We are approaching a mental health crisis of epic proportions as billions of people worldwide struggle with anxiety, fear, and depression. COVID-19 exacerbated the problem, and unchecked brain monitoring will make it even worse. Job strain is strongly linked to depression and anxiety, the risk of ulcers, cardiovascular death, and even suicidal thoughts.[62] All of which may convince even the most determined employers to give up on their efforts to monitor our attention. Many, ironically, may then fall straight into the arms of neurotech companies that offer seemingly more benign applications of neurotechnology—for example, to help employees improve their mental well-being.

The Newest Employee "Benefit": Brain Wellness

In June of 2021, the dating app company Bumble closed its offices for a week to give its seven hundred employees time to recover from their collective burnout.[63] Bumble had had a busy year, what with its stock market debut and skyrocketing growth in new users.[64] CEO Whitney Wolfe Herd was concerned about her workforce's mounting stress.[65]

At thirty-one, Wolfe Herd had recently become the youngest woman to take a company public in the United States. As she rang the Nasdaq bell, she made known her goal of making the internet "a kinder, more accountable place," an attitude reflected throughout Bumble's headquarters with wall mirrors emblazoned with messages like YOU LOOK BEE-AUTIFUL HONEY and light switches telling people to SHINE BRIGHT LIKE A DIAMOND. Bumble's employees are allowed to decide their own working hours, so long as they get their work done.[66]

While weeklong breaks may be unusual, stressed-out employees are the new normal. In an episode of the hit television series *The Office*, Dwight, a mid-level employee, set fire to the office in a misguided attempt to educate his coworkers about fire safety. Stress levels reached an all-time high as his colleagues tried to cope with the aftermath and Dwight tried to make amends. It gave us a chance to laugh at an all-too-serious and ubiquitous concern.

LinkedIn recently surveyed three thousand people in the United

States and found that 66 percent of them experience the "Sunday scaries"—feeling stressed out and anxious on Sunday nights before the workweek kicks off. Forty-one percent of respondents said their Sunday night unease had gotten worse since the COVID-19 pandemic. Generation Z was even worse off, with a whopping 78 percent of them experiencing scaries. Unreasonable workloads, poor compensation, conflict with coworkers, the looming fear of being fired or laid off, and a lack of advancement opportunities are stressing them out.[67]

The impacts show up everywhere, with one very high-profile example being Antonio Horta-Osorio—the then relatively new chief executive of Lloyds Banking Group—who was forced to take an eight-week leave of absence in 2011. "I was very mindful that the bank was in a very weak position to face adversity. It was a problem that was going around my mind constantly, which led me to sleep less and less. And the less and less sleep progressively led me to exhaustion, and then to not sleeping at all which was a form of torture, so I had to address it and I did," he explained.[68] This led him to reevaluate the mental health of the bank's sixty-five thousand employees and introduce a company-wide mental health awareness and corporate wellness programs.

Corporate wellness programs have scored some notable successes, helping employees lead healthier lives by eating better, exercising more, quitting smoking, limiting their alcohol intake, and having regular physicals. But stress remains a serious problem, undercutting corporate success by eroding creativity and productivity and raising health-care costs.[69] This is why Aetna Insurance started investing heavily in a corporate wellness program called Mindfulness at Work more than a decade ago. Employees who participated in the program, which included mindfulness training, a dedicated "calming" space for employees, and regular speaker series, enjoyed substantial reductions in stress.[70] Aetna rolled it out worldwide, and more than twenty thousand of its employees now participate in it.[71]

These programs foreshadowed the inevitable shift to wellness programs that directly target employees' brains. Like the program by Nationwide Mutual Insurance, with its benefits provider Optum-Health and its partner Brain Resource, to give employees access to a web-based tool called MyBrainSolutions, which uses brain training games to enhance their positivity, emotional resilience, and decrease their stress. More than five hundred employees enrolled within the first week, giving Nationwide baseline brain assessments of its employees.[72] Employees have continued taking brain assessments over time, unaware of the longitudinal data about their brains Nationwide could collect along the way. Brain Resource has since forged similar partnerships with other large companies, including Cisco, AstraZeneca, and Accenture.

As the reliability, availability, and costs of EEG devices to address stress improve, they are increasingly becoming part of these wellness programs. In January 2020, Morneau Shepell, a provider of technology-enabled human resources services, began a partnership with InteraXon to pilot the use of Muse headbands in corporate wellness programs.[73] Companies that sign up for InteraXon's program are given EEG devices for all their employees. The employees participate in live or remote-training sessions for meditation with a member of the InteraXon team. The program is split into managing thoughts and decreasing stress, which, as InteraXon's CEO and founder Ariel Garten claims, is "exactly what people need now."[74]

Having adopted a similar program for her own employees, Garten firmly believes that brain-wellness programs improve well-being. Employees at one company—their largest American client—became much more open about traditionally taboo topics in the workplace after enrolling in the program, she says—mental health, anxiety, personal challenges, nothing was off the table. "It allows people to open up and actually be more honest about what's going on and find real solutions for it," she claims.[75]

Offering the promise of decreased costs from absenteeism and mental health issues, Morneau hopes to roll out this program to its twenty-four thousand client organizations in 162 different countries. Though they also assure employees that their data will remain confidential, corporations that adopt the program as their own will not be bound by any such assurances.[76]

Garten herself well understands the potential risks of employers tracking employees' brains. In 2013, when she and I first met, she was the leader of the International Centre for Brainwave Ethics, an organization that carried out research on ethics, policy, legality, and data integrity in brain wave technology. After learning about our shared interest in advancing the ethical progress of neurotechnology, she invited me to join her newly launched Center for the Responsible Use of Brainwave Technology (CeReB), which was developing best practices. In 2014, CeReB issued its first major report, cautioning about the stakes in workplace settings, particularly when hiring, promotion, or firing are tied to brain data. It concluded that "a significant ethical burden is placed on organizations wishing to utilize brainwave technology in a commercial context to clearly and transparently identify the uses of the technology and the resulting data, educate users about the potential for downstream involuntary disclosure of sensitive information, ensure that consent and opt-out provisions are clearly identified and supported (without penalty), and protect the data at a level commensurate with highly sensitive personal information."[77] Unfortunately, without the broad industry funding and engagement that it needs, CeReB has been mostly dormant in the years since.

Given Garten's early engagement on these issues, I was surprised to see that InteraXon puts the burden of safeguarding employees on employees. InteraXon's privacy policy explains that its Muse Connect App allows the company "to monitor . . . the Muse sessions of one or more people using Muse devices in conjunction with the Muse Mobile application." While participants must "provide consent," once they do, InteraXon can share the employee's user activity data, processed data, and personally identifying information with their employer. InteraXon doesn't "control the Observer's use of that data," they explain. Instead, employees should "review the privacy policy of the Observer."[78]

InteraXon's policy makes no mention of sharing the raw brain wave data it collects from its users, which could limit the potential misuse of brain wave data by employers as long as InteraXon is their supplier. Hopefully that will remain true as InteraXon builds out its research program with Cambridge Brain Sciences, Hatch, and others.[79]

Countless other neurotech companies, from Emotiv to Thought Beanie, are similarly integrating EEG into existing corporate wellness programs.[80] Thought Beanie's privacy policy offers no insights whatsoever into what data they collect from users, nor what they share with employers, while Emotiv proudly touts the availability of raw brain wave data to participating employers advertising that enterprises can "access raw EEG data and Cognitive Stress & Auditory attention performance metrics."[81]

Wellness programs offer important benefits, but these data policies underscore the less visible and more dangerous privacy risks they pose.[82] Most are exempt from traditional health data privacy regulations. The Health Insurance Portability and Accountability Act of 1996 (known as HIPAA), for example, protects individuals' identifiable health information in the United States. But it doesn't apply when employers offer workplace wellness programs directly rather than in connection with their group health plans. Without laws preventing them from using the data collected through wellness programs, employers can mine the data they collect, which many do with abandon.[83] Employers learn everything that employees share in the questionnaires and online surveys they complete as part of these programs, from what prescription drugs they use to whether they voted and when they stopped filling their birth control prescriptions.[84] And employers are using that data in increasingly more intrusive ways.

Take the phone call that Chris Zubko received three weeks after he returned to work after undergoing triple bypass surgery. "Man! I notice your steps have picked up," his boss said. "You used to be under two thousand, now you're over six thousand. Two times you worked out this week. Good!"[85] Chris's boss had been monitoring his recovery through the company-installed app on his phone. The call was well intended, but it's exactly the kind of intrusion Lee Tien, a senior staff attorney at the Electronic Frontier Foundation, finds worrisome. "The more employers known about their employees' lives . . . the more potential control or effects they have on their lives," she explained.[86]

The risks extend well beyond the workplace, as wellness program vendors sell the data they collect from the programs to third-party data brokers, who use it to assess creditworthiness, set insurance premiums, and target individuals for advertisements.[87] The algorithms used by data brokers sometimes inaccurately interpret the data, mislabeling individuals as suffering from depression, mental illnesses, or drug or alcohol dependency, with predictably serious consequences.[88] A recent report from colleagues of mine at Duke examined the practices of ten major data brokers, and revealed the highly sensitive data they hold is being used in increasingly more concerning ways to exploit consumers; discriminate against individuals in housing, insurance, and education; and even target vulnerable veterans for scams.[89]

These data-scraping practices are particularly pernicious when employees' brains are involved. To realize the benefits and minimize the risks of these new programs, employees should have control over the data that is collected and be given the tools they need to act on whatever insights they supply. Employers should agree to receive only aggregate and de-identified employee brain data, which they could use to improve working conditions. Such aggregated data might show, for example, that certain activities or environments are associated with increased stress.[90] EEG and deep-learning neural networks can recognize the signs of stress with more than 85 percent accuracy.[91] An increase in beta waves, for example, is associated with rising stress levels and anxiety.[92] If employers act on such information to improve overall workplace conditions for employees, both would benefit.

But if the data is exploited for profit, wellness programs could become the most insidious form of workplace surveillance yet. That's why it's crucial that we start to develop robust norms for brain monitoring. The aim is for all of us to reap the benefits of neurotechnology, while minimizing their novel risks.

Charting New Workplace Norms

Under certain circumstances and with the right controls, surveillance technology has proven itself to be invaluable to corporations, their employees, and the public writ large. At best, it can be used to protect the public from the hazards of exhausted heavy equipment operators, to identify and address harassment and bullying at work, and to root out bias and discrimination in hiring, promotions, and firing.[93] The data it generates can lead to important insights that challenge outdated and detrimental assumptions.[94] It's legally required in the financial industry to prevent insider trading. But it can also be used to control employees, to discriminate against them, and to chill workers' ability to unionize.[95]

Companies must adopt clear workplace policies, and governments must enact laws and regulations that explicitly spell out employees' rights. This should begin with following the three-part test in international human rights law for determining whether a relative right like mental privacy can be limited: (1) legality, (2) necessity, and (3) proportionality. In other words, the human right to mental privacy would prohibit unauthorized access to employees' neural data without an explicit law justifying doing so, based on a compelling societal interest in so doing, in a manner that is narrowly tailored to balance the interests of the employee with those of society. This requires an explicit statutory framework for implementing neurotechnology in the workplace for specified purposes. Only when there is a bona fide need—a necessity—for employer-to-access employee brain data—such as to monitor fatigue in a commercial driver, or attention in an air traffic controller—should employers ask to do so. Even then, they should be limited to collecting data for the specific purpose it serves, and employers should be prohibited from mining brain data for other insights about the employee to ensure that the impact on employee privacy is proportional to the societal benefit it affords.

Neurotechnology should be a tool of empowerment, not a means of control; any law authorizing the use of neurotechnology in the workplace should require employers to provide clear notice in their employee handbooks and company policies about how they intend to use the neural data collected, as well as any insights they glean from it.

Most critically, we should prefer, if not require, that employees' raw brain data be stored on their own devices, and not on device manufacturers', software companies', or employers' servers.[96] Data should be regularly overwritten rather than stored indefinitely, as well.

Imagine the future of work when brain monitoring becomes more ubiquitous if these laws and norms are not in place. After a banner year at the company, division manager Sue calls employee Pat to offer her a contract renewal with a 2 percent pay raise. Sue knows the company could easily afford and would be willing to pay up to 10 percent to retain her but hopes Pat will take less. Pat takes Sue's call using her company-issued in-ear EEG earbuds. Pat keeps her voice even throughout the call so as not to give away her emotions and promises to follow up with Sue the next day.

All the while, Sue has been watching Pat's brain activity and decoding her emotional reaction to the news. Pat's brain activity revealed joyfulness upon learning of the 2 percent pay raise and remained joyful throughout the day.[97]

The next day, Pat calls Sue and says that she was hoping for a bigger raise. But Sue can't be bluffed; she *knows* that Pat was happy with the 2 percent raise; moreover, she now sees that Pat is fearful as she makes her request for a bigger one. Sue responds that 2 percent is the best the company can do, and Pat accepts the offer. Pat's attempt to negotiate a better salary was over before it began. Even the staunchest freedom-of-contract libertarian would question the fairness of this negotiation.

Although this is a hypothetical scenario, informational asymmetry often puts employees at a disadvantage in negotiations.[98] The asymmetry between the information that NBA team owners had about the revenues generated by games, and that the players had access to, for example, improved the owners' bargaining position during lockout negotiations in 2011.[99] Insider knowledge of research budgets has allowed unscrupulous traders to realize substantial stock gains,[100] and firm-specific knowledge has led lenders to change how they structure corporate loans.[101] When the person you are negotiating with can look inside your brain, you don't stand a chance of getting a fair bargain.

Employers could frustrate labor freedom in other ways too. Neurotechnology can't decode imagined speech just yet with any real accuracy.[102] But it can already tell us such things as when amorous feelings are brewing between employees, what a person's politics are, or whether someone is suffering from cognitive decline, and use it to fire them proactively without revealing the cause.[103] It can lead the company to look into a breach of company policy they wouldn't otherwise have suspected, and use the information selectively against employees who they suspect are planning to start a union. As neurotechnology hardware and software improve, employers might find it worthwhile to decode the vowels, letters, numbers, and words an employee is imagining,[104] or their unspoken yes and no responses to questions.[105]

Absent assurances about what brain data an employer is collecting and why, employees will be less likely to organize, as they may believe that even thinking about collective action could get them fired. Someday employers could nip labor organizing in the bud by monitoring employees' growing brain synchrony. One recent study tracked the EEG signals of high school students over a semester and found that their brain activity became more synchronized as they focused on collective tasks.[106] In other words, just by looking at patterns of brain activity across employees, it might be possible to tell who is planning something together like organizing a union. Those who are less engaged with the group can similarly be identified by their lower brain-to-brain synchrony.[107]

The use of surveillance technology to interfere with labor organizing is not just hypothetical. Stacy Mitchell describes how corporations use surveillance technology to monitor "potential for unionization," and as tools of "intimidation to stop workers from organizing and standing up."[108] In 2019, Google employees accused management of using a browser extension in their calendar application to ferret out future protests, claiming it "would automatically report staffers who create a calendar event with more than 10 rooms or 100 participants."[109]

Employers do this because employees represented by a union demand the opportunity to bargain over the terms of their employment, including the use of new surveillance technology.[110] As already mentioned, this is what happened at Rio Tinto's Hail Creek mine in 2015, when the union learned that SmartCaps were on the way.

Even nonunion employees can organize for change. More than twenty thousand Google employees and contractors around the world walked out of Google offices on November 1, 2019, after *The New York Times* reported that Google had paid Android cofounder Andy Rubin $90 million, despite allegations by a subordinate, which he denied, that he had coerced her into having oral sex in a hotel room.[111] Google ultimately agreed to several of their demands.

We can defend against potential abuses by invoking mental privacy as the basis for limiting the brain data that employers can collect to narrowly tailored purposes, like monitoring fatigue, an automatic unconscious process of the brain.[112] SmartCap does this proactively, overwriting and discarding the raw brain wave data it collects once it has been processed through its proprietary algorithms.[113] This makes SmartCap more privacy enhancing than most traditional surveillance technologies.

That's good news, but we should remember that other neurotechnology companies can and will make different choices unless laws and norms implementing the right to mental privacy prohibit them from doing so. Even when such laws are in place, employers will gain insights from neurotechnology that they can easily misuse. As Diane Jurgens, the technology officer of the mining giant BHP Billiton (which uses SmartCap to monitor drivers of its four-hundred-ton vehicles at its copper mine in Chile) put it, "You can't fool this cap because it's watching your brain waves, not looking at your eyes."[114]

But in doing so, employers invade the right of employees to define the terms of their own vulnerability. They bypass employees' mental privacy right to decide with whom they will share intimate details about themselves, and when and how they will do so.

Giving employees the right to audit brain data can help to build trust and ensure that only relevant and legitimate brain data is being collected. It also provides a check on the quality of the data being collected and an opportunity for employees to challenge invalid interpretations. If brain data is used to justify sanctions—for example, demoting or terminating a driver who repeatedly drives fatigued—employers should provide clear notice of their intent to use the data in that manner.[115]

What's at stake with the use of brain wearables at work is more than the safety, productivity, and stress of employees—it's the dignity of workers, and the future of work itself. The principle of cognitive liberty won't exclude employees from agreeing to use brain wearables at work. But those wearables can threaten mental privacy, labor freedom, and freedom of contract.

The future is already here; tens of thousands of workers are using these devices, and as neurotechnology becomes more normalized, the risk of misuse will only rise. It is past time that those laws, norms, and expectations be explicitly defined.

3

Big Brother Is Listening

In November 2021, Chinese tennis star Peng Shuai took to Weibo to make a startling accusation of sexual assault against the former vice premier Zhang Gaoli. Within twenty minutes, she was silenced by Chinese censors.[1] Soon after, all mentions of her name were purged from Chinese social media.[2] Civil rights activists, tennis luminaries like Naomi Osaka and Serena Williams, and *New York Times* investigative journalists were vocal in their support for her; within weeks, the Women's Tennis Association (WTA) had suspended all its tournaments in China. "If powerful people can suppress the voices of women and sweep allegations of sexual assault under the rug," declared WTA president and CEO Steve Simon, "then the basis on which the WTA was founded—equality for women—would suffer an immense attack."[3]

China reacted to the mounting pressure by circulating a screenshot of an email allegedly authored by Shuai in which she disclaimed the allegation.[4] They also promoted photographs from a video call between Shuai and the International Olympic Committee, which were widely derided as "stage-managed appearances" orchestrated by the government to make it seem like Shuai was well.[5] Protestors at the 2022 Wimbledon games wore shirts asking WHERE IS PENG SHUAI? to reflect lingering concerns that Shuai continues to be censored.[6]

Although the Chinese constitution ostensibly secures the right to freedom of expression, the Chinese government regularly censors or prohibits speech it believes is offensive or undermining to the Communist Party.[7] China is hardly unique in this respect—other examples abound, from North Korea to Iran and Russia. Within hours of its invasion of Ukraine, Vladimir Putin yanked independent news agencies off the air, blocked radio shows, including the Voice of America and Radio Free Europe, cut off access to websites and social media, and restricted tens of millions of Russians' use of Twitter.[8] In *On Liberty*, John Stuart Mill called attention to the insidious dangers posed by such governmental actions, arguing that humans can only "make some approach to knowing the whole of a subject . . . by hearing what can be said about it by persons of every variety of opinion, and studying all modes in which it can be looked at by every character of mind. No wise man ever acquired his wisdom in any mode but this; nor is it the nature of human intellect to become wise in any other manner."[9] Using censorship, law, and brute force, governments like those of China and Russia chill free speech to prevent their people from learning the truth about them.[10]

Repressive moves like those are terrifying enough. Even scarier is the fact that neurotechnology allows governments to move beyond the surveillance of what people are writing, saying, or doing to their very thoughts. It has already begun.[11]

In the People's Republic of China, Deayea, a Shanghai-based technology company, has revealed that conductors of the busiest high-speed rail line in the world—the Beijing–Shanghai line—have EEG devices embedded in the brims of their hats.[12] So do workers at utilities like Hangzhou Zhongheng Electric and the State Grid Zhejiang Electric Power.

It isn't just state employees who are having their brains monitored, nor is monitoring limited to fatigue or emotional distraction. In 2019, the *Wall Street Journal* reported that a primary school in Jinhua required its fifth-grade students to wear EEG headsets, which fed data to their teachers, parents, and the state. The US-based manufacturer and supplier of the devices, BrainCo, had shipped more than twenty thousand of them to China already.[13] About an inch wide and made of black plastic, the Focus 1 (or Fu Si) headsets are worn across students' foreheads. A light in the middle blazes red, yellow, or blue to signal the student's engagement.[14] More intensive brain wave data is sent in real time to the teacher's computer, whose software generates real-time alerts about students' attention levels.

The teachers overseeing the program believed that brain monitoring substantially improved their students' engagement. One student agreed, saying he had "become more attentive in class. All of my assignments come back with perfect grades."[15] Other students are less sanguine, having been punished by their parents for their low attention scores.

The story quickly went viral on social media, with photos and pictures of the students wearing the EEG devices circulating on Weibo. Chinese officials quickly quashed all discussion of the program within its borders, using its censorship machinery to scrub any mention of it from social media.[16]

Other governments are investing in brain biometrics, using brain activity to authenticate individuals, purportedly to improve border security and personal authentication. The collection of brain biometrics poses a much more profound threat to liberty than even the use of facial recognition technology, as it puts neurotechnology on a collision course with freedom of thought, a right that is already very much under siege.

Obviously not all government investment in brain research and neurotechnology is bad. They can and do fund and enable major research with an aim to alleviate human suffering.

Governments' Big Bets on the Brains

Disorders of the brain are the leading cause of disability and the second leading cause of death worldwide, and they have risen dramatically over the past thirty years.[17] So, in April 2013, when President Barack Obama unveiled the BRAIN Initiative (Brain Research Through Advancing Innovative Neurotechnologies) to give scientists "the tools they need to get a dynamic picture of the brain in action and better understand how we think and how we learn and how we remember,"[18] it received widespread and universally positive media attention. The initial $100 million Obama earmarked has grown over the past decade to nearly $5 billion under presidents Trump and Biden.[19] Among the work it has funded is research by Professor Edward Chang and his team at the University of California, San Francisco, that has yielded breakthroughs in "speech neuroprosthesis," tools that translate brain signals into text, allowing paralyzed people with anarthria to communicate.[20]

In 2016, leading scientific academies around the world called on governments to work together to carry out and fund basic research targeted at diagnosing, preventing, and treating disorders of the brain.[21] Governments heeded the call, and created an International Brain Initiative with representatives from some of the world's major brain research projects, to encourage global research collaborations and shared data standards and findings. [22]

Along with the scientific research, there's been a greater focus on the ethical implications of brain research. Shortly after the launch of the US BRAIN Initiative, President Obama asked his Presidential Commission for the Study of Bioethical Issues (of which I was a member) to consider "the potential implications of the discoveries that we expect will flow from studies of the brain . . . questions, for example, relating to privacy, personal agency, and moral responsibility for one's actions; questions about stigmatization and discrimination based on neurological measures of intelligence or other traits; and questions about the appropriate use of neuroscience in the criminal-justice system, among others."[23] In our two-volume report, *Gray Matters*,[24] we recommended that governments prioritize funding for research on brain health while supporting efforts to increase the public understanding of neuroscience and its applications. In addition, we called for equitable access to neurotechnology, offered targeted guidance on brain research for people with impaired consent, and pressed for a deeper understanding of the use of neuroscience in the legal system.

Since then, the International Brain Initiative has called upon neuroscientists to take a moral stance to guide the responsible progress of neuroscience research.[25] A moral stance that ought to include advocacy for the protection of freedom of thought.

The Forgotten Right to Freedom of Thought

In their majority opinion in the 1942 case *Jones v. City of Opelika*, Supreme Court justices Murphy, Black, Douglas, and Chief Justice Stone articulated the widely shared belief that freedom of thought is absolute, but even "the most tyrannical government is powerless to control the inward workings of the mind." Now with advances in neurotechnology we can no longer make that claim, and so we're left to confront a serious threat to our freedom with outdated legal precedents to guide us.

Freedom of thought is explicitly protected as an absolute human right in Article 18 of the Universal Declaration of Human Rights, Article 18(1) of the International Covenant on Civil and Political Rights (ICCPR) (the multilateral treaty that commits states to respect the civil and political rights of individuals),[26] and Article 9 of the European Convention on Human Rights.[27] That means, unlike mental privacy that balances the privacy rights of the individual against societal interests on a case-by-case basis, the freedom of thought of an individual should never be violated in the name of the common good.[28] Until recently, the right to freedom of thought has almost entirely applied to the protection of religious freedoms and has been undertheorized in other contexts.

In the United States, the First Amendment of the Constitution has long been understood to protect freedom of thought as a precursor to freedom of speech, which implicitly includes the right to read what one wants, think what one thinks, and not be forced to speak words or sentiments that are antithetical to our beliefs.[29] This is why state governments cannot, for example, require schoolchildren to recite the Pledge of Allegiance or be forced to recite rote facts they do not wish to say.[30] But because state regulation of *thought* has arisen so rarely, there is little to no guidance on what constitutes thought or what state actions would violate that freedom.

Now that artificial intelligence technology can increasingly infer our thoughts from our digital activities and neurotechnology can decode our emotions and one day soon perhaps even our thoughts, scholars and human rights advocates are calling for updates to international conventions.

These concerns are what prompted the UN special rapporteur for freedom of thought and religion, Dr. Ahmed Shaheed, and his team to reach out to me in 2021, to discuss a report he was preparing for the United Nations General Assembly on freedom of thought. Dr. Shaheed talked with me about how narrowly the right has been applied until now.[31] And how the use of neurotechnology by individuals, corporations, and governments could pose new threats to freedom of thought.

When I discuss "thought" as a legal interest or concern, I include in it all the ideas, reactions, reflections, images, memories, and ruminations we turn over in our minds.[32] But there is very little scientific or philosophical consensus around what "thought" should or does include. Other scholars, for example, consider the subconscious processes of the mind and one's automatic reflexes to be "thought." Jan Christoph Bublitz, a legal scholar at the University of Hamburg who has focused extensively on the issue, says that while the "*forum internum* is protected unconditionally, it is less clear what this inviolable sphere comprises."[33] Does it include all mental states? Does it include emotions and dreams? Or just "thought" in the stricter sense? Bublitz recognizes the urgency of deciding this issue, since neurotechnology can soon be used to "surmount the natural boundaries of the mind."[34]

I worry that if we include too much within the legal definition of "thought," we may interfere with ordinary human interactions. Human beings try to read one another's minds all the time. One of the earliest skills we develop during childhood is what is known as theory of mind—trying to predict what another person is thinking. If we go too far and define freedom of thought as the prohibition of every attempt to read each other's minds, we risk criminalizing human understanding.

This is why it's useful to conceptualize cognitive liberty as a bundle of rights that includes freedom of thought, the right to self-determination, and the right to mental privacy over our thoughts *and* mental processes. As an absolute human right, freedom of thought protects against governmental intrusions into our conscious thought and memories, while other aspects of brains and mental processes should be protected by mental privacy and an individual right to self-determination. This allows for the space that we need for innovation, the deployment of human empathy, and allowing some violations of mental privacy based on necessity and proportionality when societal interests call for doing so (e.g., monitoring a train engineer's brain for signs of fatigue).

When Dr. Shaheed presented his report to the UN General Assembly in October 2021, he asked the assembly to "further clarify the freedom's scope and content" while offering a framework in which to do so.[35] That framework includes the right not to reveal one's thoughts, not to be penalized for one's thoughts, and not to have one's thoughts manipulated.[36]

"Passthoughts" as a Gateway to Brain Surveillance

Assume that Meta, Google, Microsoft, and other big tech companies soon have their way, and neural interface devices replace keyboards and mice. In that likely future, a large segment of the population will routinely wear neural devices like NextSense's biosensing EEG earbuds, which are designed to be worn twenty-four hours a day. With widescale adoption of wearable neurotechnology, adding our brain activity to nationwide identification systems is a near-term reality.[37]

One of the most extraordinary discoveries of modern neuroscience is the uniqueness of each person's functional brain connectome (its physical wiring), especially in the brain areas devoted to thinking or remembering something.[38] Because of this, algorithms can be used to analyze our brain activity and extract features that are both unique to each person and stable over time.[39] How your brain responds to a song or an image, for example, is highly dependent upon your prior experiences. The unique brain patterns you generate could be used to authenticate your identity.[40]

Nationwide identification systems vary by country but generally involve the assignment of unique identification numbers, which can be used for border checks, employment screenings, health-care delivery, or to interact with security systems.[41] These ID numbers are stored in centralized government databases along with other significant personal data, including birth date and place, height, weight, eye color, address, and other information.[42] Most identification systems have long included at least one piece of biometric data, the static photo used in passports and driver's licenses. But governments are quickly moving toward more expansive biometric features that include the brain.[43]

Biometric characteristics are special because they are highly distinctive and have little to no overlap between individuals. As the artificial intelligence algorithms powering biometric systems have become more powerful, they can identify unique features in the eyes and the face, or even in a person's behavior.[44] Brain-based biometric authentication has security advantages over other biometric data because it is concealed, dynamic, nonstationary, and incredibly complex.[45]

The promise of greater security has led countries to invest heavily in biometric authentication. China has an extensive nationwide biometric database that includes DNA samples, and it also makes widespread use of facial recognition technology.[46] Chinese authorities in the Xinjiang Uyghur Autonomous Region have conducted mass collections of biometric data from the Uyghur people and used it for targeted discrimination.[47]

The United States has also massively expanded its collection of biometric data. A recent report by the US Government Accountability Office detailed at least eighteen different federal agencies that have some kind of facial recognition program in place.[48] US Customs and Border Protection includes facial recognition as part of its preboarding screening process,[49] and an executive order signed by President Trump in 2017 required the United States' top twenty airports to implement biometric screening on incoming international passengers.[50]

Increasingly, governments are investing in developing brain biometric measurements. The US Department of Defense recently funded SPARK Neuro, a New York–based company that has been working on a biometric system that combines EEG brain wave data, changes in sweat gland activity, facial recognition, eye tracking, and even functional near-infrared spectrometry brain imaging (fNIRS), a particularly promising (if expensive) technology for brain authentication, since it is wearable, can be used to monitor individuals over time, can be used indoors or outdoors while a person is moving or at rest, and can be used on infants and children.[51] China has been funneling substantial investments into EEG and fNIRS as well.

For biometric features to be successfully used for authentication, they must have universality, permanence, uniqueness, and be secure against fraud. Over time, static biometrics like facial IDs and fingerprints have become prone to spoofing. Functional biometrics, such as brain activity, are less prone to attack. That feature has motivated researchers like Jinani Sooriyaarachchi and her colleagues in Australia to develop scalable brain-based authentication systems. In one of their most recent studies, they recruited twenty volunteers and asked them to listen to both a popular English song and their own favorite song while their brain wave activity was recorded with a four-channel (an electrode capturing brain wave activity is called a channel) Muse headset. Afterward, the researchers analyzed their recorded brain wave activity using an artificial-intelligence classifier algorithm. Remarkably, they achieved 98.39 percent accuracy in identifying the correct participant when they listened to the familiar song, and a 99.46 percent accuracy when they listened to their favorite song.[52]

Using an eight-channel EEG headset on thirty research subjects, another group achieved a similar 98 percent accuracy in authenticating participants by their brain wave data after they'd looked at novel images. It might not even take eight or even four electrodes to achieve the same result. Even with just a single-channel EEG headset, researchers have achieved 99 percent accuracy in distinguishing between participants when they performed the same mental tasks.[53]

Most of these studies had a small number of participants; it is not yet clear if neural signatures will be as accurate at scale, when billions rather than dozens of people must be authenticated. EEG is inherently noisy—meaning the signals the electrodes pick up can come from eye-blinking or other movement, which can make it hard to tell the difference between brain activity or interference. But researchers have made substantial progress in developing pattern classifiers that filter noise, allowing them to discriminate between individuals based on their resting-state EEG brain wave activity and when performing tasks.[54] As noted previously, EEG devices have been used to recover sensitive information from a person's brain, such as their PIN codes,[55] and their political and religious ideologies.[56] Obviously, this poses clear risks to our digital and physical security.

Governments can already tap our phone conversations and snoop on us digitally. Will they similarly tap our brain activity data without our knowledge or consent? Will they deploy AI programs to search our brains for terrorist plots? Will they gather neural data to make inferences about individuals' political beliefs to predict and prevent peaceful protests? China is reportedly already doing so.

These and other questions underscore the need to update our right to freedom of thought. We must take steps now to ensure that brain-based authentication will not become a back door to spying on our thinking. Which is no small challenge if brain biometrics increasingly depend upon decoding our functional brain activity, as this requires decoding our thoughts. It's possible that the only way forward is to limit the data that governments can obtain from our brain to *interpretations* of our brain activity, and not the raw brain data itself. An algorithmic interpretation of that data that simply confirms "this is a match" or "this is not a match" would limit the further processing of brain activity data by government actors. To cede more may well put our cognitive liberty at perilous risk.

But even limited interpretations of brain activity may not suffice to preserve our sphere of thinking freely. Whatever security advantages we gain from brain-based biometrics may be outweighed by the impact of giving governments the power to track our brains.

The Chilling Effects of Brain Surveillance

The chilling effects of government surveillance have been extensively documented. Dr. Elizabeth Stoycheff, a professor of communications at Wayne State University, studies the ways that mass surveillance affects people's behavior.[57] In one study, Stoycheff created a baseline psychological profile of research participants based on their surveyed ideological beliefs, personality traits, and online activity. Then she subtly reminded a random subset of those participants that they were subjects of mass government surveillance. Afterward, all the participants were shown a made-up newspaper headline which stated that the United States had undertaken airstrikes against the Islamic State in Iraq and were asked their opinion about it, including how they thought other Americans would feel and whether they would be willing to voice their own opinions in public. The participants who had been primed to think about mass government surveillance were significantly more reluctant to share their nonconforming views, even when their personality profiles predicted otherwise. This underscores the significant impact of self-censorship in response to surveillance and reinforces decades of research on the "spiral of silence" that was first identified in 1974 by the German political scientist Elisabeth Noelle-Neumann—the phenomenon in which the perception that one's opinion is unpopular makes one reluctant to express it.[58]

What most dismayed Stoycheff was people's cavalier dismissal of surveillance. "So many people I've talked with say they don't care about online surveillance because they don't break any laws and don't have anything to hide. I find these rationales deeply troubling," she mused. "It concerns me that surveillance seems to be enabling a culture of self-censorship because it further disenfranchises minority groups. It is difficult to protect and extend the rights of these vulnerable populations when their voices aren't part of the discussion. Democracy thrives on a diversity of ideas and self-censorship starves it. Shifting this discussion so Americans understand that civil liberties are as fundamental to the country's long-term well-being as thwarting very rare terrorist attacks is a necessary move."[59]

Freedom of thought is at the heart of those civil liberties; without it, the diversity of ideas necessary for human flourishing is silenced. Just as surveillance chills people from sharing their nonconforming views, thought surveillance will inevitably lead people to attempt thought modification—trying to silence their inner voices, risking a dangerous spiral that ends with the suppression of even their innermost views. Making it of paramount importance that we prohibit governments' surveillance of thought.[60]

Government surveillance of brain activity will inevitably push us toward greater conformity. With greater conformity comes a passive acceptance of authority and authoritarianism, either out of fear or in hopes of appearing cooperative, even when that conflicts with one's own moral compass.[61] Children are particularly susceptible to pressure to conform and so are even likelier to try to redirect divergent thinking for fear of being ostracized. Many of the worst atrocities are "crimes of obedience": acts carried out in response to orders from authority that violate legal and social norms.[62]

To know the difference between right and wrong, and to decide for ourselves what that is, we must have the freedom to think critically about the world around us.[63] Freedom of thought guarantees us a private space to think and self-reflect, where we are free from fear of reprisal. This gives us the wherewithal to reject orders that we know are wrong.

This freedom is critical for all of us, not just great thinkers. John Stuart Mill made this point eloquently in *On Liberty*, arguing that "it is as much, and even more indispensable to enable average human beings to attain the mental stature which they are capable of."[64] When we have the freedom to think, we can decide for ourselves whether we want to be angry about a setback or an insult from another; we can investigate our feelings and align our instincts with our self-identity.[65] But we can do so only in a mental space that is free from government surveillance.

A Thought Experiment on Thought Crime

On the first day of my criminal law class for first-year law students, I assign the classic science fiction film, *Minority Report*, which is based loosely on a Philip K. Dick short story. Set in Washington, DC, and Northern Virginia in 2054, the film follows the pursuits of the Precrime unit, an elite group of police who, relying on intelligence from "Precog" psychics, arrest potential criminals before they commit their crimes. The upside is that the Precrime unit has eliminated all planned murders in its jurisdiction. The downside is that they are suppressing a minority view among the Precogs, that a suspect could still choose a different future.

After discussing the film, I ask my students to consider an alternative future in which we could use neurotechnology to anticipate future crimes. If new technology revealed that someone was contemplating murder, should we arrest them? Even when I head off the question of the reliability of the technology at the pass by stipulating that it is bulletproof, my students invariably come down strongly against arresting suspects based on their thoughts alone. They worry about the loss of our collective sense of personal security and liberty. They may believe that law enforcement has a role to play in deterring crime, some even citing examples of teenagers intercepted by the police on their way to commit a crime, potentially saving them from a lifetime behind bars. But they believe equally strongly that we should give people every opportunity to renounce their criminal intent and should limit police powers to interfere with individual liberties only when a person actually does something wrong.

Now that I have their attention, I reveal the startling truth: governments are already using neurotechnology to detect people's thoughts and memories. And they are prosecuting and convicting them based on what they discover.

Interrogating the Brain for Crime

When investigative journalists David Kocieniewski and Peter Robinson broke the story about the ties between Donald Trump's incoming national security advisor, Michael Flynn, and a company that sells brain wave technology to governments worldwide, surprisingly few people noticed.[66] Serving alongside Flynn on Brainwave Science's board of directors was Subu Kota, a software engineer who had pleaded guilty to selling highly sensitive defense technology to the KGB during the Cold War.[67]

Brainwave Science sells a technology called iCognative, which can extract information from people's brains. Among its customers are the Bangladeshi defense forces as well as several Middle Eastern governments.[68] Following some successful experiments at the Dubai Police Academy, Emirati authorities have recently deployed the technology in real murder investigations. At least two cases have successfully been prosecuted.[69]

In one case, the police were investigating a killing at a warehouse.

Suspecting that an employee was involved, they forced the warehouse workers to don EEG headsets and showed them images of the crime. Purportedly, a photo of the murder weapon triggered a characteristic "recognition" pattern in one of the employee's brains (the P300 wave), while none of the other employees showed a similar response. Confronted with that evidence, the suspect confessed, revealing details that only the guilty party could have known.[70]

First reported in a series of experiments published in *Science* magazine in 1965, a P300 wave is an event-related potential (ERP) measurement of brain activity,[71] or more simply, an automatic brain response that happens when we encounter some specific sensory, cognitive, or motor event. Research subjects wore EEG headsets while listening to sounds and flashes of light that were presented in pairs, with a three-to-five-second delay between them. In the first experiment, light followed light, and sound followed sound. In the second experiment, light was followed randomly by either light or sound, and vice versa. By comparing the participants' brain activity between the first and second experiments, the researchers discovered a consistent brain response that occurred about three hundred milliseconds after the target stimulus. But its amplitude differed, depending on whether the participant was certain about the target stimulus that would follow or was uncertain about what came next.

Larry Farwell, an independent researcher, wondered whether those signals of certainty and uncertainty might be useful in police interrogations.[72] In 1991, in an article titled "The Truth Will Out," he reported that he had successfully used the P300 waveform to detect concealed information in the brain. Farwell first trained research subjects on a fake espionage scenario, and then had them act it out. Each participant went to a specific location to meet a person, exchanged a password, then asked that person for a file, who passed it to them.

The next day, the research subjects wore the EEG headsets while a series of questions or "probes"—two-word descriptive phrases about the crime scene—were flashed on a computer screen.[73] Whereas a traditional polygraph test asks subjects a series of yes and no questions and then analyzes their physiological responses to infer if they are telling the truth, Farwell examined subjects' brain responses to phrases like "green hat" (relevant to the person they exchanged the file with), "Tim Howe" (relevant), and "Ship Plans" (relevant) versus "brown shoes" (irrelevant), "Ray Snell" (irrelevant), or "Plane Plans" (irrelevant). The subjects' P300 brain signals revealed recognition of the relevant details of the crime but not the irrelevant ones. Because people can't control their unconscious brain responses, Farwell argued, it would be much harder to "beat" his test.[74]

In November 2001, when the country was obsessed with terrorism, Farwell was profiled in *Time* magazine as one of America's one hundred "breakthrough" scientific innovators.[75] The "brain-fingerprinting" technique he developed, the article read, would allow interrogators "to determine if a subject is familiar with anything from a phone number to an al-Qaeda code word." The CIA funded his research and police departments enlisted him to help solve difficult murder cases. But for all the popular enthusiasm, scientists couldn't replicate Farwell's findings and eventually rejected his methods as little more than clever marketing.

Despite the decades that have passed since, most scientists remain appropriately skeptical about the scientific validity of brain fingerprinting for criminal interrogation. The design of the key phrases or images to test people with requires substantial expertise and involves a degree of subjectivity, which some scientists have called more art than science. Although Farwell became one of the original founders of Brainwave Science (a partnership that ended with a series of lawsuits over disputed patents), and Robin Palmer, a former criminal defense lawyer from South Africa, has independently validated some of his findings,[76] the otherwise broad and seemingly warranted derision of the technique by the scientific community makes it all the more troubling that governments worldwide have used it as often and as recently as they have.

Some of the earliest efforts were at the behest of criminal defendants, who hoped the test would validate their claims of innocence.[77] In 1978, Terry Harrington was convicted for the murder of John Schweer, a nighttime security guard in Iowa.[78] Harrington appealed his conviction and was eventually granted a new trial in 2001, and the state of Iowa decided not to try him again. At one point during the appeals process, he consulted with Dr. Farwell, who developed a series of unique probes based on previously undisclosed evidence from police files and Harrington's alibi. In his expert report, Farwell claimed that Harrington did not recognize any of the crime details from the police files, but that his brain did register recognition of the alibi probes.[79] Although the Iowa Supreme Court ultimately decided the case on other grounds, by admitting the expert testimony at trial, it paved the way for the future introduction of brain-fingerprinting evidence in criminal cases in Iowa, as the judge found it admissible.

More commonly, it is the police who seek to use the technology. James B. Grinder was the primary suspect in the brutal rape and murder of Julie Helton in 1984, but there wasn't enough forensic evidence to connect him to her death.[80] The police asked Grinder if he would be willing to submit to brain fingerprinting. Certain he could beat the test, Grinder agreed. When the results revealed brain activity consistent with his recognition of the crime scene, Grinder pleaded guilty and was sentenced to life in prison without the possibility of parole.[81] While there was still a chance that a judge would find the evidence scientifically inadmissible and exclude it, Grinder didn't want to risk facing the death penalty.

Despite that success, the practice has yet to become widespread in the United States, not because of concerns about freedom of thought but because criminal defendants refuse to submit to it voluntarily. As Brainwave Science promotes its adoption worldwide, we can expect its use to continue to spread. Police in India have been using brain-fingerprinting technology since at least 2003. Singapore's police services purchased brain-fingerprinting technology in 2013; the Florida State police signed a contract to use it in 2014. And Australian counterterrorism authorities are looking into the potential use of brain fingerprinting to determine whether individuals returning from war zones who claim to have been carrying out humanitarian work were in fact involved in the conflicts.[82]

Brain-fingerprinting technology using P300 is just one of several approaches to probing the brain. A scientifically promising approach uses a different ERP brain response called the N400. You could show a suspect a series of faces that includes their suspected coconspirators. Their N400 brain signals would be more negative for "incongruent" faces that didn't belong than for "congruent" faces that did. Similarly, you could pair words together like "body" and "lake" versus "body" and "basement" to try to find out where a murder victim's corpse is hidden.[83] DARPA's Neural Evidence Aggregation Tool program is exploring the promise of N400 signals to interrogate the brain for "congruent" and "incongruent" facts.

Other approaches use functional magnetic resonance imaging (fMRI) to analyze truthfulness. The premise behind fMRI lie detection is that a few key areas of the brain are more active when a person is lying than when they are telling the truth.[84] Two companies—Cephos Corporation in Tyngsboro, Massachusetts, and No Lie MRI in San Diego, California—marketed this technology for a time. But it is cumbersome and unvalidated, particularly for "real-world" and high-stakes lies compared to low-stakes lying in controlled laboratory settings.

While various protections afforded to criminal defendants could be brought to bear against the use of these techniques in some countries, most fail to give adequate legal protection against their use. Even the Constitution's Fourth Amendment protection against unreasonable searches and seizures and the Fifth Amendment privilege against self-incrimination are unlikely to keep the government from probing our brains. Until now, these protections have been held inapplicable in certain contexts, such as when information was obtained from a device manufacturer rather than through a search of a person's body or home. Similarly, if the police obtain incriminating evidence from a person's brain, courts may interpret that as physical evidence and thus not subject to the privilege against self-incrimination. Laws in other countries are likely to be interpreted similarly.

We can see a trend in how police are using evidence from other wearable devices to predict the direction this will go. Consider the case of Richard Dabate, charged with murdering his wife in 2015. Dabate told Connecticut police that a masked intruder shot her before tying him up. But the Fitbit device she was wearing revealed that she had still been moving for an hour after Dabate said she was killed.[85] Objective evidence from a Fitbit device also became critical in the investigation of the 2016 murder of Nicole VanderHeyden. Her boyfriend, Doug Detrie, became a prime suspect in the killing after police found blood on the floor of their garage and in Nicole's car. But Detrie claimed he was asleep at the time of her killing and had awoken only once to check on their six-month-old baby. Detrie had been wearing a Fitbit device, and its data corroborated his story. Another man was ultimately arrested for the murder.[86]

Will we come to regard our brain data in the same way we do Fitbit and GPS data? One of the distinguishing features of brain-fingerprinting technology as opposed to traditional forensic evidence is that you have to question a person or show them images or other stimuli to provoke a brain-based response. This means the person is provoked to create incriminating evidence against themselves, which is more likely to violate existing rights like the right against self-incrimination than the passive creation of evidence by existing fitness trackers. But marry a future in which we wear neurotechnology all the time with the ability of the police to obtain its data and decode it, and the passive tracking of our brain activity may also become a reality that can be used as evidence against us in a criminal investigation. Shouldn't freedom of thought include protection against having the government use our memories and silent utterances against us?

The Right Not to Reveal One's Thoughts

In 1890, when future Supreme Court justice Louis Brandeis was thirty-four years old, he worried about the "numerous mechanical devices" that would soon allow "what is whispered in the closet" to be "proclaimed from the house-tops."[87] It wasn't neurotechnology that was keeping him up at night. Brandeis was fretting about portable cameras and celebrity journalism. He and his law partner Samuel Warren were so concerned about the implications of these advances that they penned a groundbreaking essay that argued for a legal right to privacy. Published in the *Harvard Law Review*, it would become the foundation for many of our modern-day privacy protections.[88] While Brandeis and Warren couldn't have anticipated governments' use of neural data, they nevertheless argued for the protection of our "thoughts, sentiments and emotions." Nearly a century later, in *Stanley v. Georgia*, the US Supreme Court formally embraced their view that the Constitution secures us against "governmental intrusion into one's privacy and control of one's thoughts."[89]

After he joined the Supreme Court, Brandeis wrote several prescient opinions about modern technology's potential impingements on our privacy. In his 1928 dissenting opinion in *Olmstead v. United States*, he predicted that "The progress of science in furnishing the Government with means of espionage is not likely to stop with wire-tapping. . . . Ways may someday be developed by which the Government, without removing papers from secret drawers, can reproduce them in court, and by which it will be enabled to expose to a jury the most intimate occurrences of the home."[90] Twenty years after *Olmstead*, the United Nations General Assembly adopted the Universal Declaration of Human Rights, including its Article 18 protections of freedom of thought. By 1966, that right was enshrined in the International Covenant on Civil and Political Rights.

But no one, not even Brandeis, anticipated that governments could one day tap directly into our minds, deciphering the emotions, sentiments, and even unuttered speech that they detect. Nor could they have imagined that we, as a society, might one day acquiesce to such intrusions. But seismic shifts in science and technology are often followed by seismic shifts in our understandings of rights.

In his 2021 report to the UN General Assembly, Special Rapporteur Dr. Shaheed underscored the importance of freedom of thought to one's "*forum internum*—a person's inner sanctum (mind) where mental faculties are developed, exercised, and defined." These faculties are critical to our ability to "perceive truth, to choose freely and to exist." Or, as John Stuart Mill cautioned in *On Liberty*, without freedom of thought, we "dare not follow out any bold, vigorous, independent train of thought, lest it should land [us] in something which would admit of being considered irreligious or immoral."[91] No one can ever become a great thinker if they cannot follow their thoughts to wherever they may lead.

Just as we must take steps to ensure neurotechnology isn't used as a back door to spying on our thinking, we must also take steps to ensure it isn't used to weaponize our thought against us.

Once our brains can be probed for crime scene details, it isn't hard to imagine some governments going further and punishing people for their ways of thinking, especially if they are thinking about organizing to overthrow a tyrannic regime. Will George Orwell's dystopian vision of thought crime become a modern-day reality?

The risks that accompany the benefits of using our brains as tools of law enforcement go to the very heart of being a thinking human being; the urgency of our need to update our definitions of freedom of thought cannot be overemphasized. The ICCPR's General Comment 22, last updated in 1993, emphasizes the broad scope of freedom of thought but focuses on freedom of religion and belief. We now need to update our understanding of the international human right to freedom of thought to include our right to think without threat of thought surveillance, and without our thoughts being used against us—whether in a criminal proceeding or any other context. In the chapters ahead, we will explore other aspects of freedom of thought that need updating, including the right not to have our thoughts manipulated or our very capacity for thought assaulted.[92]

To realize the true promise of neurotechnology, we must learn how to harness it for ourselves. We can't do that unless we have unfettered access to our own data, an issue we turn our attention to next.

4

Know Thyself

Many scouts believed that Justin Skyler Fields would be a top five pick in the 2021 NFL draft and very likely the first quarterback selected. Fields was an all-star at Ohio State University, a finalist for the Heisman Trophy, and had been named the 2019 Big Ten Offensive Player of the Year. But just eight days before the draft, Justin threw a monkey wrench into the league's evaluation of him when he went public with his epilepsy diagnosis.[1]

When Justin was in high school, he woke up disoriented and confused in an ambulance, later learning that he'd had his first epileptic seizure. Fortunately, and like most people with epilepsy, he responded well to medication. In fact, he never missed a college game. "I mean, it's pretty simple for me to manage it," he explained. "I just have to take three to four pills a night, every night. It's nothing crazy. It's a thing that's been there for the past seven or eight years, so I'm used to it."[2]

With neurotechnology, Justin could soon learn about an impending seizure before it happens through real-time alerts on his smartphone or just-in-time calls from his doctor's office. He could track his brain activity over time and share it with team scouts to show that his condition is well-managed. But whether individuals *should* have direct access to their own brain data or rely on their doctors instead to learn about their brain functioning is a question we will soon have to address as neurotechnology becomes more widespread. After Justin's announcement, he fell to the eleventh pick overall and the fourth quarterback off the board, selected by the Chicago Bears. The Bears' general manager, Ryan Pace, said that his team's medical evaluators had gotten in touch with their peers at Ohio State and decided that they were "completely fine with it," having "dealt with something similar in the past with different players over the years."[3] But other teams without similar experience with epilepsy may have been more hesitant to draft him.

It's not surprising that a star football player would suffer from epilepsy. The most common serious brain disorder, epilepsy affects about fifty million people worldwide.[4] Epileptic seizures are associated with short bursts of abnormal electrical signals in the brain. A so-called grand mal seizure renders a person unconscious and sends them into convulsions, but not all seizures are that dramatic; some can cause a brief feeling of disorientation, while in other cases a person may appear normal and awake but is actually in a trance. Any kind of seizure could be extremely dangerous for someone driving a car—or playing pro football.[5]

That's the kind of informational uncertainty that the teams likely grappled with after Justin's bombshell announcement. His draft stock likely fell because not all seizures are as well controlled as his. It was no secret to the league that in 2007, the Baltimore Ravens' standout safety player Samari Rolle had suffered three major seizures that kept him out of six games.[6] An adjustment to Rolle's medication allowed him to play again but losing a star quarterback for several games could be devastating for a team's playoff potential.[7]

About a third of adults and about 20 to 25 percent of children with epilepsy do not respond well to conventional anti-seizure medications. If two different medications are tried and both fail to keep them seizure-free, their epilepsy is classified as drug-resistant. There aren't a lot of great options for people in that category. They could undergo surgery to remove the part of the brain that is causing the seizures, be put on a special diet, take immunotherapy drugs with serious side effects, or rely on benzodiazepines that become increasingly less effective as their brains develop a tolerance to them.[8]

Consumer neurotechnology may change that picture. EEG has been effectively used to diagnose and manage patients with epilepsy, and machine learning algorithms can use brain data to discern neural signatures of imminent seizures.[9] Researchers from Ben-Gurion University of the Negev in Israel have already developed a wearable EEG device called the Epiness, which they claim can predict seizures up to an hour before they occur and send a warning to a smartphone.[10] Clinical trials are underway.

Fitted with a device like Epiness, a person with epilepsy could drive to work confident that they won't have an accident on the way. A sufferer of drug-resistant epilepsy could take benzodiazepines on an as-needed basis to mitigate an impending seizure, preventing the risk of tolerance that develops with continual use.

The potential of these devices is astonishing but having direct access to our brain data is threatening to a deeply entrenched system in which information about our bodies is filtered to us through intermediaries. Regulators, physicians, and even most bioethicists(!) still believe that consumers lack sufficient medical literacy to correctly interpret data about their own brains and bodies and are thus prone to making dangerous choices. The solution, they say, is to restrict how much information consumer devices and apps can display.

Regulators worldwide will classify Epiness and devices like it as medical devices, subject to stringent premarket and post-market controls. That means extensive premarket testing on the safety and efficacy of the product before it can be marketed, and the requirement that only experts can have direct access to its software. As consumers, we will of course want reassurance that Epiness and similar devices are accurate. But if they do what their makers claim, should an expert be the one to decide whether and how we can use them?

Many doctors, regulators, and cultures believe that information about our health and medical treatments should be filtered and contextualized, and that to do otherwise causes unnecessary and avoidable harm, while others believe that the patient's right to informed consent requires sharing with them more fully. It's time we critically examine these claims. Can having access to accurate information about our brains actually *harm* us?

From Meditation to Alzheimer's Disease

Dr. Richard Davidson, a neuroscientist at the University of Wisconsin–Madison, had been meditating for more than forty years when he redirected his research to the effects of meditation on the brain. This shift in his work occurred at a meeting with the Dalai Lama, who wondered at the focus of his research: "You've been using the tools of modern neuroscience to study depression, and anxiety, and fear. Why can't you use these same tools to study kindness and compassion?"[11] This question would lead him on a journey to find out just how much we can learn from EEG.

Davidson and his colleagues began by flying Yongey Mingyur Rinpoche, a Tibetan monk and a master of the practice of mindfulness, from Kathmandu, Nepal, to Madison. Using a sophisticated clinical-grade EEG setup, the researchers monitored Rinpoche's brain wave activity as he alternated between meditations on compassion and thirty-second rests.

As Mingyur began his first meditation, the researchers saw a huge burst of electrical activity on the screen.[12] At first, they assumed that Mingyur must have moved, because movement creates artifacts that often confound accurate EEG detection.[13] But they soon discovered that this burst of electrical activity occurred every time Mingyur meditated on compassion. When they repeated the experiments on twenty-one other Buddhist monks, they discovered the same thing. Davidson and his colleagues proved not only that EEG can detect the neural changes that occur during meditation, but that meditation has powerful and lasting effects on the brain.[14]

Other researchers have conducted similar experiments since. In one study, researchers recorded the brain waves of Buddhist monks in the northeastern region of Thailand via a Muse EEG headband. The monks answered questions, read books, and meditated, and based on their brain wave activity alone, the researchers could tell which activity they were engaged in.[15]

As related in the story that opens this book, today, companies like InteraXon are marketing EEG headsets to allow consumers to "see" if their subjective experience of meditation aligns with the brain wave data their devices decipher. While some studies give reasons for caution about the accuracy of the data the devices report, others have shown that people enjoy significant increases in mindfulness after using them, leading to reductions in stress, heart rate, and an improved sense of well-being.[16] That monitoring could also reveal more devastating information, for example, that a brain tumor has started to grow.

Diffuse gliomas are the most common and aggressive form of brain cancer. There are very few effective treatments; the relative five-year survival rate for diffuse midline glioma is less than 50 percent. While many factors affect a person's prognosis, early detection, especially in younger patients, is critical, because it allows for intervention before it spreads. But traditional screening strategies generally fail at early detection. Case reports and serial brain scans show that small brain lesions can evolve into established disease in as few as sixty-eight days.[17]

Herein lies the promise and peril of portable EEG devices for individuals. About 68 to 85 percent of brain tumor patients have abnormal EEG profiles when they are first diagnosed. We aren't there yet, and we have a way to go to even learn if this is possible, but someday, regular monitoring with consumer devices may allow us to catch glioblastomas in their earliest and most treatable stages.[18] They might be able to catch a host of other diseases, as well. The South Korean biomedical startup iMediSync is already marketing an EEG device that can detect the mild cognitive impairment that heralds early Alzheimer's dementia with 90 percent accuracy as well as other neuropsychiatric disorders, such as Parkinson's disease, traumatic brain injury, PTSD, ADHD, depression, and more.[19] The data the device captures is sent directly to physicians and specialists, who then inform the patients of the results. Having completed its clinical trials in South Korea, iMediSync has received regulatory approval from the Ministry of Food and Drug Safety (MFDS).[20]

But is it in fact better to route all "bad news" through an expert like a physician? Would we be better off routing it through someone trained in empathy and compassion? Or letting patients decide whether they want to learn about it themselves first, through a connected app and smart device? And do some patients fare better if they *don't* know about the diseases that may be lurking in their brains?

A Good Lie

When I left North Carolina for college, I doubted I would ever move back to the South. But when I was getting ready to attend graduate school, my mother was diagnosed with myelofibrosis, a very rare and progressive form of bone-marrow cancer, and I knew that I had to be close to her. Which is why I received my JD, MA, and PhD from Duke.

Since then, I have thrown myself into learning as much about my mother's illness as I can. I've read everything I can find about the disease, its available treatments, and my mother's prognosis. This is how I navigate uncertainty in life—by arming myself with knowledge. With this knowledge, I have been able to advocate for my mother, joining her for many of her medical appointments and even reviewing her regular blood test results.

My father, a retired physician, has also been deeply involved in her care, and it is through our mutual work together that I have learned something the medical journals and textbooks leave out: how to tell a good lie.

When my mother's blood cell counts started to plummet, a sign of disease progression, I started to explain this to my mother, drawing a sharp look from my father. He took me aside and admonished me that a physician learns when to share, but also what *not* to share with their patients. What good would it do to tell her the implications of those results? "Medicine is about much more than treating a patient's condition," he said. "It is about learning to shield a patient from information that won't serve them well."

I was taken aback. But when I researched the question, I learned that his view is not uncommon. Like many physicians of his generation and cultural heritage, my father believes that it is not only unnecessary to share some information with patients, but that doing so may *harm* them.

Eric Topol, the founder and director of the Scripps Research Translational Institute, and author of best-selling books on the future of medicine, writes elegantly about the history of this perspective in his 2015 book *The Patient Will See You Now*, tracing the idea that physicians should patronize their patients and decide whether and how much information to disclose to them from 2600 B.C. to the present. Topol frames our present-day informational asymmetry as one in which "Doctors have all the data, information, and knowledge. Patients can remain passive or ignorant of their medical information, or, if they choose to be active, they typically have to call repeatedly or beg to get their data."[21] The reason for this, he explains, is because their doctors are "trying to 'protect' the patient by not disclosing adverse, anxiety-provoking information."[22]

Underlying this approach is the assumption that only an experienced physician, with years of training and unique knowledge of the patient's medical history, has the context and knowledge to know what information to share and when. And that they are in the best position to balance when and whether the benefits of a device or treatment outweigh its risks.[23]

This approach presumes that some kinds of information—like information about your brain functioning—are so arcane that the average layman can only react emotionally to it; that they lack the objectivity and hence the autonomy that they need to confront it directly and responsibly. Having access to truthful information about your brain functioning may cause you unnecessary anxiety or psychological distress, the theory goes, or cause you to pursue harmful medical treatments to mitigate risks that you don't understand.[24] Or you might waste precious medical resources by seeking further evaluations that you don't need.

As a legal ethicist, I struggle factually and philosophically with these claims. First, the empirical research shows people adapt quickly to information about themselves. Worse still, the information denied is critical to making choices about how they will live their lives. And yet, as a daughter who deeply loves her mother, I can't help but notice that my mother doesn't ask very many questions about her prognosis. She seems to actually do better when she *believes* that her disease is under control. In a way, she is asserting a right of self-determination; for her, it may be better not to know, or she may not want to know. But does her subjective experience bear on whether *you or your mother* should have the right to informational self-determination?

Consider whether you would want to know if you were about to have an epileptic seizure. If so, would you prefer to hear about it through a phone call from your physician, or a real-time alert on your smartphone, particularly if the availability of the latter allowed you to drive or play sports or do other things that are part of a full life? How about whether you are meditating properly? Doctor or cell phone app? What about if a brain tumor has taken root or if you are suffering from the earliest stages of Alzheimer's disease? Does your preference for what you learn and how you learn it vary by the kind of information that is in play?

Consumer neurotechnology will force you to grapple with these choices. Regular monitoring of your brain wave activity could allow you to "see" that you have achieved a meditative state, but also that you may be showing signs of early Alzheimer's disease. Does the right to informational self-determination include the right to decide how you learn about either one?

To Know or Not to Know

Questions about whether and when to share "bad" news remind me of a recent film, *The Farewell*, which follows a Chinese family who decide not to tell their grandmother Nai Nai that she is dying of lung cancer.[25] Using the pretense of a cousin's wedding, the family gathers in China to say their final goodbyes.

The main character, Billi, struggles to reconcile her family's deception of Nai Nai with her own personal guilt in doing so, underscoring the narrative tension the film draws between contrasting moral frameworks—Western individualism and Eastern collectivism.

To Western audiences more accustomed to the idea that individual autonomy reigns supreme, the deception of Nai Nai might seem incontrovertibly wrong. But that deception is both familiar and demanded by many Eastern cultures. In countries like Iran, China, Singapore, Japan, and Lebanon, and even in immigrant subpopulations within the United States like my parents', the practice is common. Many Iranian and Chinese families object to sharing a "bad" diagnosis or prognosis with a patient, and some experts recommend that the wishes of the family be respected. The deception is both well meaning and consistent with an ethical norm of non-maleficence—avoiding unnecessary harms to individuals. These cultures believe that individuals may die faster when they know they have cancer, because of the fear it instills.[26]

A recent survey of Chinese doctors found that 98 percent of them would tell family members about a cancer diagnosis before telling the patient, and 82 percent would follow the family's wishes as to whether the patient should be told.[27] While the Western approach to disclosure of information is now different, it hasn't been that way for very long. A 1961 study in Chicago surveyed doctors on this same question. Ninety percent said they wouldn't inform a cancer patient of their diagnosis, and that they would deliberately mislead them to protect them.[28]

Recent studies have shown that knowing about a terminal diagnosis doesn't shorten a person's lifespan—and that knowing better enables a person to make treatment decisions, end-of-life care choices, and personal choices about their remaining life and estate.[29] In fact, a recent meta-analysis of twenty-three different studies on the issue—which reviewed 11,740 records—found no evidence that uninformed cancer patients had a better quality of life or fewer disease-related symptoms than informed cancer patients. In fact, informed patients showed *greater* vitality than uninformed ones.[30]

The same finding—that people do better when they know the truth about their condition—has emerged in research studies again and again, even when the information is as dire as having a high risk of developing Alzheimer's disease. In the case of 23andMe discussed in chapter 1, the FDA expressed concern that people would react poorly to learning exactly this kind of information. The empirical evidence shows otherwise. For most people, knowing about their risk of Alzheimer's disease (based on apolipoprotein E genotyping) does not increase their risk for psychological harm. Often it leads them to have more positive feelings about the risk assessment experience. Of course, most people *are* distressed in the short term. But this temporary distress doesn't lead to an increase in anxiety or depression overall. And the same people who proactively sought out their risk of developing Alzheimer's disease are more likely to use that knowledge to adopt healthier lifestyle habits and engage in better long-term planning.[31]

The weight of this empirical research, together with shifting norms, has created a growing trend of more honest communication with patients.[32] Informed consent laws in the United States, Europe, and China, as well as international human rights laws, require full and complete disclosure of information to patients about their conditions. These laws have evolved over the past fifty years from a focus on what a reasonable physician would find necessary to disclose to a patient to what a reasonable patient would want to know, based on their right of self-determination.[33] This legal trend follows a demographic one in which younger people expect—nay, demand—access to information.[34] Generation X and millennials are far less likely than the generations before them to trust health-care providers, and far more likely to trust self-directed information.

While we can't yet be sure how people will react to what they learn about themselves from consumer neurotechnology, we can see that "good lies," even about serious diseases, are hardly a better alternative. Consumers are more likely to benefit from informational self-determination over their own brain data. Of course, not everyone will choose to know. My mother says she would not. But a right to informational self-determination gives a person that choice.

Headwinds Against Change

Less than a decade ago, it would have been unthinkable for a person with a healthy brain to track their own brain activity. If you were curious about your brain, you would have to see a specialist, who would order imaging only if he or she believed it was medically warranted. Further, the tests would provide only a snapshot of the brain at a given moment.

But thanks to the proliferation of new technology, a cultural shift is underway; consumer-based self-quantification is fundamentally changing how we learn about our biological selves. If you walk into your local pharmacy, you can buy home testing kits and devices that track your heart rate, sleep patterns, blood sugar, moods, and even your attention span.[35]

With every step forward, there are myriad attempts to turn back the clock, reinserting "experts" between consumers and their data.

If you suspect you have a fever, more than likely you'll just grab a thermometer from your medicine cabinet and take your temperature. But in 1867, when England's Sir Thomas Allbutt invented the first practical clinical thermometer, only a physician could read the results.[36] Fifty years ago, regulators and physicians believed that laywomen lacked the competence to self-administer pregnancy tests and react to them appropriately.[37] British and Canadian regulators didn't approve do-it-yourself home pregnancy tests until 1971,[38] while the United States' FDA held out until 1976.[39]

Rapid HIV home tests followed this now familiar arc. The OraQuick In-Home HIV Test allows individuals to test themselves or a partner (or would-be partner!) for HIV and have an answer in about twenty minutes. But during the HIV/AIDS crisis in the late 1980s, regulators worldwide took a hard stance against HIV home testing, dogmatic in their view that consumers needed those results to be contextualized by a physician lest they react in "hysterical or irrational ways, such as committing suicide."[40] The FDA waited twenty-five years before it approved the first home test kit for HIV in 2012,[41] and it wasn't until 2015 that many other countries followed suit, including England, Scotland, Wales,[42] and China.[43]

In each of these cases, consumers ultimately received unfettered access to the new technologies—a reason for optimism when it comes to neurotechnology with appropriate safeguards for our mental privacy. But when we look at the recent debates around direct-to-consumer (DTC) genetic testing technology, the future looks much less rosy.

One Small Step Forward, One Giant Leap Back

In 2006, after attending a scientific conference, the US biologist and entrepreneur Linda Avey took an audacious gamble. Now that it was possible to extract DNA from saliva as well as blood, she could massively scale the amount of genetic material available to scientists, allowing researchers to learn much more about genetic contributions to human traits and diseases, while at the same time empowering people to learn more about their own genomes. All she had to do was invite people to spit into a plastic tube.[44]

Linda pitched the idea to her former boss at Perlegen Sciences, Paul Cusenza, who was just as excited as she was. They worked furiously over the following months to develop a business idea and started shopping it to potential investors. When they pitched it to entrepreneur Anne Wojcicki, she became the third cofounder of the company that became 23andMe. When the marketing team asked the founders for one word that described their product, Linda replied, "Audacious!"[45] 23andMe boldly challenged the expert intermediary system by marketing its product directly to consumers. At first, it offered limited genetic sequencing paired with regularly updated reports about variations a person had at points along their genome, and what those variations meant. They started by reporting on fourteen different conditions and soon expanded to 254 disease and health predispositions.[46]

As an early adopter of the 23andMe service, I enjoyed a much richer ecosystem of informational offerings than is now available, including my unsurprising risk of rheumatoid arthritis (my father has it), scoliosis, and Hodgkin's lymphoma, and my decreased risk of Alzheimer's disease, stomach cancer, and celiac disease. I received my carrier risk status for more than fifty different diseases, my likely responses to twenty-five different drugs, and information about nearly sixty different genetic correlations to traits such as my likelihood of avoiding repeat errors and degree of caffeine consumption. 23andMe also specified their level of confidence in my report, quantified my personal risk level relative to the population, and provided me with a genotype summary and all the relevant citations to learn more. The confidence level on many of those early reports was low and the research quite thin, so I didn't know what to make of that data. I was bummed to learn I had a higher risk of rheumatoid arthritis, but I haven't lost any sleep over the information, and so far, haven't developed it. Equipped with the data and the research that informed their insights, I could decide what *I* thought about the information they reported.

But there was a storm coming. Linda left the company in 2009.

Shortly afterward, motivated by the sector's increasingly health-related claims, the FDA started to increase its oversight of DTC genomic testing companies,[47] warning that it could classify their services as medical devices subject to stringent premarket approval requirements that would be nearly impossible for most companies to meet.[48]

Between 2010 and 2012, the FDA ramped up the pressure. 23andMe worked with the agency to classify its product as a lower-risk medical device, which would have allowed it an easier pathway to regulatory approval.[49] Then its general counsel left, and the lone remaining cofounder, Anne Wojcicki, inexplicably stopped responding to the FDA.[50] At the same time, the company launched a major national TV campaign, touting the purported health benefits of the product—including your risk of celiac disease—and encouraging consumers to use the service to discover their risks.[51]

We will never know what might have been if 23andMe had not ghosted the FDA. What we do know is that on November 22, 2013, 23andMe received what is known as a "warning letter" (a regulatory enforcement action against the company) that described its personal genome service (PGS) as an unapproved Class III device—meaning that it needed to go through stringent premarket safety and efficacy testing—and ordered the company to "immediately discontinue marketing" it.[52] Despite its multiyear engagement with 23andMe, the FDA said that it lacked "any assurance that the firm analytically or clinically validated the PGS for its intended uses."[53]

An earlier paragraph in that now infamous letter suggests a different reason for the FDA's action, which was concern that the average customer could overreact to some of the information they received. "For instance," the letter stated, "if the BRCA-related risk assessment for breast or ovarian cancer (like the one [Angelina] Jolie received from her physician) reports a false positive, it could lead a patient to undergo prophylactic surgery, chemoprevention, intensive screening, or other morbidity inducing actions."[54] In other words, the FDA worried that "patients relying on such tests may begin to self-manage their treatments," which raises "serious concerns . . . if test results are not adequately understood by patients or if incorrect test results are reported."[55]

My Duke colleague Misha Angrist, a bioethicist, called out the FDA for being "borderline absurd" in assuming that a woman would get a double mastectomy based on a cheaply available consumer test without following up with a physician—or that a surgeon would operate solely on that basis.[56] But the now all-too-familiar concerns about consumer health literacy had hijacked the dialogue.

23andMe tried to hold out, but on December 5, Wojcicki announced that the company would stop offering the service.[57] When she later clarified that the company would still offer existing customers their raw DNA data,[58] I rushed to download it, along with an archive of my old reports. 23andMe has since become a poster child for regulatory compliance. Whereas before, you could have access to hundreds of reports, now only a few dozen are made available to consumers.[59]

Other countries have enacted onerous laws to restrict consumer access to genetic testing in the name of consumer safety, adding layers upon layers of restrictions, including mandatory supervision of genetic testing and limitations on the way it is performed. France, Hungary, and Germany have mandated that it can be performed for health-care purposes only, with a medical prescription and undertaken by an authorized laboratory. Other countries impose mandatory genetic counseling before consumers can receive their reports. South Korea strictly limits certain kinds of genetic tests, such as those that predict physical characteristics or personality traits.[60]

Most doctors and geneticists support these restrictions, believing that the average person lacks "health literacy," which undermines their autonomy and renders any right of access to their own data meaningless.[61] In my view, the FDA's actions against 23andMe were a tragic defeat for patient empowerment and a threat to freedom of speech by restricting our free access to information.[62] Alongside other academics, like Gary Marchant, the Regents Professor of Law at Arizona State University, I have argued that consumers have a right to the information contained in their own genes and that the FDA's approach was appallingly paternalistic. The legal ethicist Barbara Evans emphasizes that the moral principles upon which bioethics rests define "autonomy as the capacity to make one's *own* decision, regardless of whether society would see that as a *good* decision. Autonomy is about the right to be wrong."[63]

In other words, while there is an important information-forcing role for regulators to play to ensure consumers receive robust safety and efficacy data about drugs and devices, it should do so to inform their choices, not prevent them from making one.

Now a growing chorus of experts are worried that consumers will misuse or misinterpret their brain data.

Anna Wexler, University of Pennsylvania professor of Medical Ethics and Health Policy, advocates for increased regulatory oversight because consumers are "left to themselves to evaluate the veracity of neuroscientific claims—a task in which very few are well trained."[64] Scholar and communications expert Caitlin Shure wrote to the FDA to call for "proactive safeguards against unsubstantiated claims by neurotechnology manufacturers" to protect "vulnerable consumers."[65] Shure argues that our "understanding of brain electrophysiology is extremely limited," and that consumers lack the "base of intuition with which to discriminate the value and validity of neurogadgets."[66]

As Jack Nicholson famously cried out in the movie *A Few Good Men*, this growing chorus of experts believe that when it comes to your own brain data, "You can't handle the truth." The FDA's regulatory approach suggests they may well agree, foreshadowing a straitened future for consumer neurotechnology applications.

The FDA has already rated EEG,[67] EMG,[68] and fNIRS products as Class II medical devices, meaning that it considers them to pose moderate to high risks to consumers, and thus are subject to specific regulatory controls rather than being treated as "generally safe" and okay to market just by notifying the FDA. While this is still less onerous than Class III devices, in that it does not require premarket safety and efficacy studies, manufacturers must file premarket submissions to the FDA to show that their devices are safe and effective or substantially equivalent to other already-approved products. Until the FDA gives a manufacturer its okay, the manufacturer cannot market or sell a new Class II device.

There is an exception in the rules that manufacturers have tried to exploit to bypass these premarket regulatory controls. The FDA does not regulate "general wellness" products designed to promote "a healthy lifestyle."[69] If a product is intended for "relaxation or stress management," "mental acuity," "concentration," or enhancing "learning capacity,"[70] it can be sold directly to consumers without premarket controls or prescriptions. But if *and only if* the product is intended solely for general wellness use.[71]

As a result, the vast majority of manufacturers limit their marketing claims to those "wellness" categories. They market products that allow you to wriggle EEG-powered cat ears with your emotions,[72] improve your golf game through neurofeedback,[73] play video games by thinking about moving around the screen,[74] hone your focus by mentally opening and closing a flower,[75] monitor your sleep activity,[76] meditate with biofeedback,[77] or navigate AR/VR[78] hands-free. But don't let this fool you into believing that your access to neurotechnology isn't at risk. Despite calls for reassurance from the industry, the FDA has *not* promised that consumer neurotechnology qualifies for this loophole. In its "general wellness" guidance document, the FDA doesn't mention a single neurotechnology device as "low-risk," and it has ignored explicit requests to do so.

And we shouldn't ignore the more worrisome message that is lurking within existing regulatory guidance. Consumer neurotechnology holds much greater promise than playing games with our minds; those devices can potentially revolutionize how we diagnose and treat neurological conditions. But when Dr. Daniel Johnston, formerly of the Pentagon's Executive Medicine Office of the Surgeon General and now the cofounder of BrainSpan, an integrated nutritional and functional brain health assessment system, asked the FDA to include terms like "cognitive/brain performance" (in addition to concentration), or "tracking brain performance" (when it mentioned "mental acuity"), or language like "detect cognitive decline or change as a result of aging or injury" within its wellness guidance documents, the agency declined to do so.[79]

When 23andMe starting advertising its product as a tool to learn about our disease risks, regulators worldwide shut it down. As consumer neurotech manufacturers begin to do the same, will they face the same fate?

Using Health Literacy as a Sword

The common thread in all these debates is whether you have a right to unmediated information about yourself.

I believe traditionalists are on the wrong side of history. The view that consumers are too health illiterate to justify self-access to their brains belittles the average person and denies them the opportunity to become more educated.

To teach a person to read, you wouldn't take away their books; you would teach them how to read them. Requiring an expert to filter all our data is like allowing only experts to read books to the illiterate out loud. Moreover, a troubling aspect of health literacy claims is the historical backdrop of using literacy as a tool to disenfranchise people. Echoes of those efforts reverberate through calls to deny individuals tracking of their own brain activity. The wrong side of history, indeed.

Using consumer neurotechnology, we may learn hopeful and fascinating and even potentially upsetting facts about ourselves. But that is no reason to suppress or limit access to that information if it is trustworthy. None of this is to deny that those experts and regulators have an important role to play when it comes to neurotechnology—we need their carrots and sticks to incentivize manufacturers to conduct robust safety and efficacy testing of their products and disclose those results to consumers. That empowers consumers, rather than disenfranchises them.

Your Right to Know Thyself

The philosopher Aristotle believed the self is in the heart, while René Descartes argued the self is embodied in a soul separate from the body.[80] But when the Yale psychologist Christina Starmans and her colleagues showed children and adults pictures of flies circling around their bodies and asked them to point to the pictures where the flies were closest to their "selves,"[81] the participants invariably pointed to the ones where the flies were closest to their eyes. Other research shows that this holds regardless of age or culture.

Our strong intuition that our sense of self is inextricably bound up with our brains makes self-access to our brain activity uniquely important to our self-determination. When Neo wanted to understand his fate in the epic science fiction movie trilogy *The Matrix*, he went to the Oracle to find out. The Oracle helped him learn that he had to discover the truth himself, by looking inward. While who we are is importantly different from the data produced by our brains and nervous systems, access to that essential information about ourselves is central to the self-reflection and self-knowledge we need to develop our own personalities—a right that has been explicitly recognized by the European Court of Human Rights (ECtHR) as part of the privacy protected in Article 8 of the European Convention on Human Rights in the case of *Satakunnan Markkinapörssi Oy and Satamedia Oy v. Finland*, where the court explained how "The protection of personal data is of fundamental importance to a person's enjoyment of his or her right to respect for private and family life," and found that "Article 8 of the Convention thus provides for the right to a form of informational self-determination, allowing individuals to rely on their right to privacy as regards data which, albeit neutral, are collected, processed and disseminated collectively . . . "[82] That approach should inform an updated understanding of the Universal

Declaration of Human Rights to include a broad international right to informational self-determination.

The German Federal Constitutional Court first defined the right to informational self-determination in a 1983 opinion that described "the authority of the individual to decide himself, on the basis of the idea of self-determination, when and within what limits information about his private life should be communicated to others."[83] A proactive aspect of informational self-determination is our right to access and record our own personal information. "This right," explains the Argentinian philosopher Gabriel Stilman, "includes the power to record or collect data" that we want to document about ourselves, including, for example, our own "identifiable brain activity."[84]

While the right to information has been most fully realized as a right to information held by governments, the special rapporteur for Freedom of Expression and Opinion has called for an expanded understanding of Article 19 of the Universal Declaration of Human Rights, the right to "freedom of opinion and expression" and to "receive and impart information and ideas through any media and regardless of frontiers" to include the right of information "held by public bodies in the broadest possible terms."[85] People should "be able access adequate, accessible and necessary information as soon as it is known."

I believe the right to obtain and record your own brain activity should be included within our understanding of the right to privacy and right to freedom of opinion and expression in international human rights law, as an essential precondition to cognitive liberty. General Comment 34, last updated in September 2011, recognizes the right of every individual "to ascertain in an intelligible form, whether, and if so, what personal data is stored in automatic data files, and for what purpose."[86] International human rights law should be interpreted to include access to brain data collected automatically by neural devices should we wish to receive that information. Without the right to informational self-determination, other important rights, like mental privacy, and freedom of thought and speech, are put at risk. How can we protect freedom of speech, if we cannot record, show, and then share information and circumstances from our own brains?[87]

Of course, our right to informational self-determination is more meaningful if the enforcement of that right includes specific provisions to ensure the information in question is accurate. And manufacturers are perversely incentivized to be the first to market—even if that means their claims are inaccurate or misleading. As a society, we should combat misleading claims by companies about their products and penalize companies that consistently produce inaccurate data.[88] To do so is to mandate a corporate duty of discovery and effective disclosure: to honestly label their products, do their part in educating consumers, and only sell products that have been vetted by objective parties to ensure that they do what they say they do. But there are far better ways of doing this than denying consumers access to these technologies. Regulatory and market mechanisms can nudge companies toward greater transparency and self-reporting. Fines and enforcement actions can be taken against companies that make false or misleading claims. And responsible companies will help to educate consumers about what their results could mean.

Linda Avey described to me her pride in the fact that 23andMe offered videos and other educational materials to consumers that could help them contextualize difficult news. Technology enables manufacturers to deliver similarly useful information, like peer-reviewed journal articles, in a just-in-time manner. To do so effectively, they must communicate using clear and effective language, delivered through media most likely to help consumers understand the information being provided to them.

Third-party solutions from other industries can serve as models. While the dietary supplement industry is largely unregulated, organizations like ConsumerLab.com have emerged to deliver high-quality information to consumers about different supplements and their effects, and conduct independent testing to empower consumers' decision-making.

Regulators have an important function to play by putting effective incentives for informational disclosure into place for drug and device manufacturers. Often market mechanisms fail to nudge manufacturers to be fully transparent about the limitation of their products. Just as nutrition labels are standardized to help consumers compare the nutritional contents of different foods, consistent labeling requirements by neurotech manufacturers would enable consumers to compare products against each other. Such labels would make details such as how many electrodes are used, what is being recorded, the error rate in detecting brain wave activity, what brain wave "bands" are being recorded, where the electrodes are placed, how accurate claims are about the associations between brain waves being detected and the function being measured. (If a device claims to detect seizures, for example, how often is a positive reading correct? How often is a seizure missed? How often is the prediction wrong? What is the data that supports those claims?)

Consumer protection agencies should continue to take action against manufacturers that make false and misleading claims, while implementing a series of carrots and sticks that would require manufacturers to avoid making false and misleading claims and to disclose information that would guide consumers in effective ways. With a recognized right to cognitive liberty, we can blaze a better path forward.

But self-determination is about more than our right to access information regarding our own brains. To define the contours of self-determination over our brains and our mental experiences, we need to decide if we can not only track our own brains but also hack them. And what rights, if any, have we against being hacked by others?

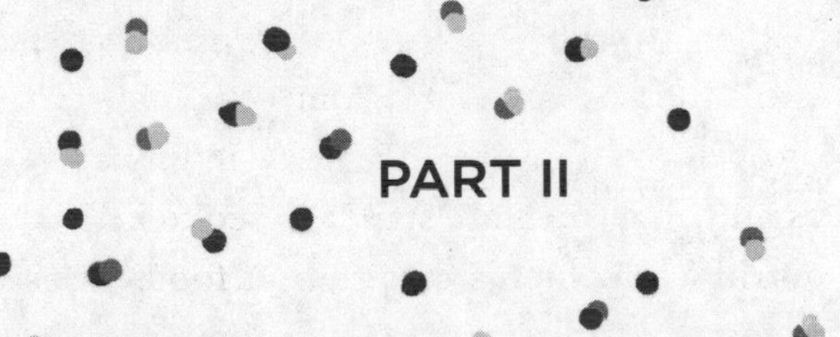

PART II

Hacking the Brain

5

Revving Up

An audience member at a recent talk approached me with a question. Her son Ethan was preparing for his SATs, and she wondered if she ought to encourage him to take an attention deficit hyperactivity disorder (ADHD) drug on test day to improve his performance. Ethan didn't have ADHD, but she knew that a lot of healthy students take ADHD medications to improve their concentration and enhance their school performance. Admission to college has become so incredibly competitive, she explained, that Ethan would need whatever help he could get.

We talked about some of the risks and benefits of ADHD drugs, including the idiosyncratic ways that people respond to them. I cautioned her that it would be particularly risky for Ethan to try a new drug for the first time on test day, because he wouldn't know how it would affect him. She nodded, concerned about that possibility. But she didn't ask me the one thing that many others may have also pondered: Would taking a prescription drug that was not medically necessary give Ethan an unfair advantage over other students whose parents may not know a friendly doctor, or have the means to pay for one? Would Ethan be cheating?

I don't know what Ethan decided to do. What I do know is that if he *did* take an ADHD drug on test day, he would have joined the legions of other healthy people who are asking their doctors to prescribe or are purchasing drugs on the black market to enhance their brains.[1] Bookstore shelves are overflowing with titles like *The Genius Within*, that describe the huge array of licit and illicit drugs, behavioral modification strategies, and physical devices that purport to unlock more of our brains' potential.[2] Consumers are game, trying out each new offering, dubious though their efficacy may be, hoping to gain a cognitive edge.

A recent study revealed that the rate at which healthy people take brain-enhancing drugs nearly tripled between 2015 to 2017, from approximately 5 percent to almost 14 percent.[3] In the United States, nearly one in five respondents had used a brain-enhancing drug. From prescription drugs for ADHD like Adderall (dextroamphetamine-amphetamine) and Ritalin (methylphenidate) to narcolepsy medications like modafinil and even illegal stimulants like cocaine, the trend is rising across individuals and populations.[4]

Ethan's profile fits squarely within the largest contingent of healthy users—eighteen-to-twenty-five-year-olds taking Adderall or related drugs. Most get them from their friends or family members who have prescriptions.[5] Whether (and if so, how much) these drugs improve their brain performance is unclear. Published studies on the value of cognitive enhancers for healthy people suffer from small sample sizes and their conclusions vary depending on the kinds of brain functions that are tested.[6] But at least some drugs seem to work. Healthy, non-sleep-deprived people who take the stimulant modafinil enjoy improved attention, executive functioning, and learning.[7] They may similarly experience improvements in memory or the rate at which they process information after taking Ritalin or other prescription stimulants.[8] Some drugs, such as those that increase acetylcholine in the brain by blocking the action of the enzyme cholinesterase, can improve memory, alertness, verbal fluency, and creative thinking.[9]

Even if Ethan opted not to take a drug, he could join the growing ranks of consumers who use dietary supplements. Because of long-lasting brain fog in people who have recovered from COVID-19, the makers of products like Prevagen, a supplement that contains vitamin D3 and apoaequorin (a protein in jellyfish); omega-3 fatty acid supplements; ginkgo biloba;[10] and nootropics like Mind Lab Pro—which all promise to improve memory, focus, and overall brain health and function—are experiencing skyrocketing demand worldwide, even though the evidence in support of their claims is modest at best and the dietary supplement industry is largely unregulated.[11]

The brain-training game industry is also booming; it was worth a whopping $8 billion in 2021.[12] These games are designed to exercise our brains, much the way we exercise our muscles.[13] Many of the techniques they incorporate can produce modest improvements in working memory or other task-specific functions, but they require substantial investments of time and money.[14] That said, some of the game makers' claims about their effectiveness have been so overblown that members of the scientific community issued a consensus statement in 2014 to caution consumers about the glaring lack of studies linking specific games to specific improvements.[15]

But for at least two of these games, that may no longer be the case. When Australian scientists weighed the scientific evidence for the efficacy of brain training games, they found that the studies backing up BrainHQ and Cognifit were well designed, controlled, and properly randomized. People who played these games for a few weeks had measurably improved processing speed, attention, memory, reasoning, and executive functioning.[16]

These effects may be even stronger when brain training is paired with the use of a consumer EEG device that enables learning reinforcement by neurofeedback. Neurofeedback gives us real-time control over our brains.[17] When we can see how our brain wave activity changes in response to certain things we do, we can change our behavior to change our brains. While these products are still in the developmental stage, the global market for neurofeedback devices is expected to reach $57 million by 2025.[18]

Yet other ways to track and hack our brains are on the horizon. Elon Musk hopes to soon begin clinical trials of "the Link" (implantable electrodes being developed by his company Neuralink) to create direct brain–computer interfaces for people with paralysis, which could one day be used to enhance healthy people as well.[19] Other major corporate players, like Synchron and Blackrock Neurotech, are even further along with their implanted neural chips.

What should we make of all of this as an industry? Is it enabling a society of cheaters? If that is the case, does that mean we are also cheating when we ingest sugar,[20] caffeine,[21] cocoa, curry powder, *Bacopa monnieri*, folic acid, and even herbal remedies, which have all been shown to boost our cognitive performance?[22] How about when we take SAT preparation classes? Music enhances our brain function, as do good nutrition and regular exercise. Is it cheating to listen to Mozart while we jog and then eat a healthy breakfast? Will cognitive enhancers further widen the gap between haves and have-nots?

Perhaps we need to think about brain enhancements in a different way.

Operation Varsity Blues

For a clear example of cheating, consider the scandal now known as Operation Varsity Blues. In March 2019, the US Department of Justice arrested more than fifty people who had participated in it, among them Hollywood actors and other wealthy parents, exam administrators, and athletic coaches from Yale, Stanford, USC, Wake Forest, and Georgetown University. Rick Singer, the scheme's organizer—the founder and CEO of the Key, a "Private Life Coaching and College Counseling Company"—collected nearly $25 million to ensure that students received preferential treatment in admissions because of falsified athletic accomplishments, and to help them cheat on their ACT and SAT exams.[23] They did this by paying doctors to diagnose the students with learning disabilities so they could qualify for special accommodations, like being given extended time on their tests. Once the accommodations were granted, they changed their testing center to one where the proctor was on Singer's payroll. On test day, either paid ringers would take the tests instead, or the proctor would feed the students correct answers. Sometimes the parents didn't tell the student about the arrangement; in those cases, the proctor would correct the exams after the students had finished.[24]

The public was rightly outraged at the spectacle of these wealthy elites rigging the system in favor of their privileged children at the cost of all the genuinely high-caliber students who lost their spots to them.[25] As the trials proceeded, the conversation shifted to the already-skewed processes of college admissions in the United States. Parents can donate huge sums of money to universities to gain a leg up for their children when they apply.[26] As one reporter put it: "Fraud and bribery are shocking, yes. But fraud and bribery's lawful cousins—legacy preferences, athletic recruitment, and other admissions practices that lower the bar for progeny of the rich and famous—are ubiquitous." [27]

The blatant cheating that was part of this scandal—passing off someone else's work as your own—is not only unethical but also illegal in many jurisdictions. As early as the 1970s, several US states passed laws against term-paper-writing services and other businesses that enable academic cheating.[28] Many states make it illegal to fraudulently obtain or even attempt to obtain any diploma, certificate, or other instrument purporting to confer any literary, scientific, professional, or other degree at any institution of higher education.

Other countries do the same. Australia has a new law in draft that tackles contract cheating by students. The law defines cheating as completing an assignment or work for a student, providing any part of a piece of work or assignment, providing answers for an examination, or sitting for an examination in the place of a student.[29] Under a similar law, six people were jailed in China for helping dozens of students cheat on China's national exam for graduate schools. The group provided the test-takers with wireless transmitters and receivers and instructed them to read the questions aloud. Using textbooks and other resources, off-site researchers sent them the correct answers.[30]

Everyone agrees that it's wrong for students to cheat. But where do we draw the line between what counts as cheating and what doesn't? Where on that spectrum should we put the use of brain-enhancing goods or services while preparing for exams? Are the advantages that they confer necessarily unfair?

At least one academic institution seems to think so.

Cognitive Enhancements on College Campuses

In 2011, in response to pressure from students, my home institution, Duke University, changed its academic integrity policy to count the use of cognitive-enhancing drugs for nonmedical reasons as cheating. Violators faced expulsion. According to Duke's policy: "Cheating is the act of wrongfully using or attempting to use unauthorized materials, information, study aids, or the ideas or work of another in order to gain an unfair advantage. It includes, but is not limited to . . . the unauthorized use of prescription medication to enhance academic performance."[31]

Notably, the university didn't consult with bioethicists or any other faculty on campus. Other universities have also struggled with what to do about the use of ADHD drugs by neurotypical students. Wesleyan University has an "Adderall clause" in its Code of Non-Academic Conduct.[32] Like Duke, it calls out nonprescription use of Adderall as academic dishonesty.[33] Even Great Britain's Academy of Medical Sciences characterizes the unauthorized use of neuroenhancers, which they called "study drugs," as cheating, very much like the use of steroids in sports.[34]

Some schools, such as Vanderbilt University and Phillips Exeter

Academy, attempt to regulate neuroenhancers in the same way that they do illegal recreational drugs. While prescription drug abuse is considered a serious-enough offense to merit expulsion, they do not list it as an academic offense.

Still other schools make it hard for students to obtain prescriptions for ADHD drugs on campus: California State University, Fresno, requires two months of testing and extensive paperwork before the student health office can diagnose ADHD. Even then, a student must sign a formal contract, agreeing to submit to drug testing, see a Fresno State therapist regularly, and not share the pills. The policy also "does not allow early refills to replace lost or stolen medication. Urine tests can be required should a university clinician suspect that a student is not taking the pills as prescribed."[35] The University of Alabama and Marist College require students to sign similar contracts, while George Mason forbids its clinicians to diagnose ADHD. William & Mary clinicians refer ADHD students to off-campus providers for their prescriptions, while Marquette's require students to sign releases allowing them to phone their parents for full medical histories. The University of Vermont will not perform diagnostic evaluations for ADHD. North Carolina State, Georgia Tech, and Penn State refuse to do evaluations for ADHD because of the volume of requests and the "time it takes to do the evaluations right."[36]

Like the students at Duke who pushed for a change in the school's academic integrity policy, the people who crafted these rules believe that taking these drugs creates an "unfair advantage for the users who are willing to break the law in order to gain an edge," and that students who take advantage of such chemical "shortcuts" have poor work ethics and are depriving themselves of the opportunity to develop stronger time-management skills.[37]

Is It Really Cheating?

While it's undoubtedly illegal to obtain prescription drugs under false pretenses, I believe that these universities get the moral calculus wrong. Improving our brains' function isn't cheating, whether we do it through studying, exercising, diet, brain-training games, devices, or neuroenhancing drugs. The gradual improvement of our brains over time is fundamental to human flourishing; it is a social good that we can and should pursue. Moreover, it is a social good that we pursue *all the time.*

To the extent that smart drugs and devices improve our focus, motivation, attention, concentration, memory, we ought to celebrate rather than prohibit them. We are all constantly striving to open our minds up to new experiences and improve our ability to learn new things. Why should we treat a new way to do so as impermissible?

Better brain functioning can make us more successful at work, enhance our earning potential, reduce our likelihood of experiencing social and economic difficulties, and improve our overall health. What's at stake is at the core of cognitive liberty—the right to self-determination over our brains and our lives.

While it's true that college admissions in the United States are becoming more competitive, and that access to and the use of brain-enhancing drugs could conceivably raise a student's chances of qualifying for a more elite school, that benefit is one that other students could equally realize should they also use those brain-enhancing goods. One person's use does not trade off with another's, just as one person's enjoyment of a concert, art installation, or good meal will not trade off with another person's enjoyment of the same. What's unethical is unfairly limiting access to them. Moreover, restricting access to cognitive enhancements is most likely to burden those who are the least well off. The haves will find a work-around to gain access—a friendly private doctor, for example—while the have-nots will continue to be left behind.

Widespread improvements in cognitive function could lead to significant benefits for society. Shouldn't we want our scientists, factory workers, truck drivers, and other members of society to operate more effectively, with better concentration and capabilities? If using brain-enhancing drugs, games, or devices enables us to study harder and longer while learning and retaining more, won't that improve our odds of solving many of the ills that plague society? Of curing diseases like cancer, now the leading cause of death? Won't it speed the development of tools that bring us closer together, solve social ills, and allow us to create more beauty in the world, while improving our overall happiness? Shouldn't we encourage, rather than quash such opportunities?

As for whether some people may have access to these drugs while others do not, that would suggest that we as a society ought to make access to those drugs, games, and devices more rather than less equitable. Not ban them or deem it cheating to use them. Making them more available is a better way to level the playing field than restricting access to those few who can game the system to obtain them or afford the high price of the black market. We certainly shouldn't criminalize the quest for cognitive flourishing.

Nor should we compare the use of smart drugs and devices to the use of steroids in sports, an analogy that has led us very far astray. Life isn't a competitive game with winners and losers and spectators on the sidelines. Brain improvement is not zero-sum. All of us stand to gain something from it.

The Misplaced Sports Analogy

By the year 2000, when she was twenty-four, everyone was talking about Marion Jones. In the three years since she'd become a full-time track athlete, she had dominated female track-and-field competitions so commandingly that most of her rivals were unknown. She had won thirty-three 100-meter races in a row, and all twenty-two of the 200-meter races she'd run. By the end of 1997, she was the world champion in the 100, and the next year she became the first woman in fifty years to win three individual events at the national championships. She'd won thrity-six out of thirty-seven sprints and jumps and was the second-fastest woman ever in the 100 and 200.[38] When she set herself the goal of winning five medals at the Sydney Olympics, all eyes were on her.

And she did: three gold and two bronze. By the end of the year 2000, she had been named the fastest woman in the world. Her sponsors—from AT&T and TAG Heuer to Nike—all touted her as "the total package."[39] When asked to explain her success, she attributed it to "50 percent talent and 50 percent just wanting it so badly . . . I train harder than I think anybody else in the world in my sport."[40]

Jones's rock-star status stood her in good stead when her then-husband C. J. Hunter tested positive for steroids in the middle of the Sydney Olympic Games. And it helped sustain her popularity when she took a year off from sports in 2003 after announcing that she was having a child with her then-boyfriend and fellow sprinter Tim Montgomery. But then the US Anti-Doping Agency began probing Jones's ties to the Bay Area Laboratory Co-operative, which had been accused of distributing steroids to numerous top athletes. Jones's boyfriend was charged with steroid use, and her ex-husband told federal officials that he had injected her with banned substances and seen her inject herself with them before, during, and after the Sydney Games.

By 2004, Jones's performance had slipped notably. Many now suspected that the source of her success hadn't been the 50 percent talent and 50 percent determination that she had touted, but her use (and likely abstinence when under scrutiny) of anabolic steroids. Jones vehemently denied it until October 2007, when she admitted in federal court that she had started using a designer steroid known as "the Clear" before the Sydney Games and had continued to use it until July 2001.[41] Following her admission, the International Olympic Committee formally stripped her of her medals and wiped her name from the record books.[42] Her fall from grace and from international fame was swift and complete.

What Makes Performance-Enhancing Drugs Count as Cheating in Sports?

Why was Jones's loss of status unambiguous? And how do we know that using "the Clear" was cheating?

The World Anti-Doping Code, which was first adopted in 2004 and that seven hundred sports organizations around the world have now signed on to, provides the rationale for anti-doping in sports. Its purpose, it claims, is to preserve "the spirit of sport," which it defines as "the pursuit of human excellence through the dedicated perfection of each person's natural talents." Since doping is unnatural, it is seen as fundamentally contrary to the spirit of sport, to the protection of athletes' health, and to their "right to compete on a doping-free level playing field."[43]

In other words, we ban certain substances from sport because as a society we have decided to *define* sport as about the celebration of natural talents, honed only through societally approved enhancements, and to protect the health of athletes by banning substances that pose risks to them. Games are self-enclosed systems; their rules and components define them. If it's in the rules, it's part of the game. If it is not, it isn't! It's just that simple.

Unlike other aspects of life, like attending school because of compulsory education laws, playing a game is entirely voluntary and cannot be forced upon a player. Anyone who plays a game voluntarily is bound to its rules. Typically, the goal of the game is for an individual or a team to win. There are preestablished conditions that must be met to do so.

There is no question that doping improves muscle strength and performance. Studies on the use of androgens show they can improve muscular strength by 5 to 20 percent, even when they are taken in much lower doses than athletes typically use.[44] The enormous rewards for winners, the ever-increasing effectiveness of the drugs, and the fact that only 10 to 15 percent of athletes are randomly tested during major competitions create tremendous incentives for their use.

Of course, we could decide to define sports in some other way, such as a celebration of steroid-enhanced talent. In 1988, the television show *Saturday Night Live* ran a comedic riff on this very idea with "the first All-Drug Olympics," in which athletes were not only allowed but encouraged to take all substances before, during, and after competitions. In this alternative world, 115 world records had already been shattered, as a Soviet weight lifter, Sergei Akmudov, who had ingested "anabolic steroids, Novocain, Nyquil, Darvon, and some sort of fish paralyzer" as well as "a few cocktails within the last hour" attempted to clean-and-jerk more than 1,500 pounds to triple the existing world record. Instead, his arms were pulled off his body and prop blood spurted everywhere.

But so far, we are sticking to the old approach. In the zero-sum games of sports there are clearly defined rules of participation.[45] Deviate from them, and you are a cheater. Of course, reasonable people could disagree as to whether the "spirt of sport" privileges those who have won the genetic lottery over those who could achieve the same effects through ingesting or injecting other agents into their bodies. Consider the role of the hormone erythropoietin in sports performance.

The Finnish skier Eero Mäntyranta dominated the sport of cross-country skiing in the 1960, 1964, 1968, and 1972 Olympics, winning seven medals in total. In his prime in 1964, he won the 15-kilometer event by 40.7 seconds, which is a margin of victory that no winner had accomplished before or since. As it turned out, Mäntyranta had a rare mutation in the gene that produces the receptor for the hormone erythropoietin (EPO).[46] When oxygen levels in the tissues of the body drop, the kidneys produce EPO, which tell the bone marrow that it needs to manufacture more red blood cells. Once oxygen levels are stabilized, EPO production is shut down and red blood cell production returns to normal. But Mäntyranta's genetic mutation turned off this feedback loop, so that his body continued to make extra red blood cells—as much as 40 to 50 percent more than the average individual. Some athletes have tried to mimic the same effect by training at high altitudes where there is less oxygen, conditioning their bodies to produce more EPO and hence more red blood cells. Others, such as the now-disgraced Lance Armstrong, took synthetically produced EPO to achieve the same effect. (Synthetic EPO has been banned since the early 1990s.)

So why was Mäntyranta celebrated as a star athlete and Lance Armstrong treated as a cheater? Because the rules we have set say that naturally occurring advantages are permissible whereas ingested, injected, or mechanical ones are not. A seven-foot-tall basketball player has an advantage over a shorter player, but the shorter player isn't allowed to use stilts to compensate. As a society, we can choose to do this in a zero-sum setting like sports where what's at stake—who wins or loses a game—is far less significant than what's at stake in brain enhancements, which is individual and societal progress.

We've also made a societal decision that the protection of athletes' health matters. When players at Princeton and Rutgers faced one another in the first college football game in 1869, neither team wore headgear. It wasn't until 1893 that protective helmets came into use. As the rules of the game evolved to allow more player-to-player contact and tackling, the players' equipment also evolved. And yet, it wasn't until 1939 that helmets became a mandatory piece of equipment in college ball.[47] Heightened concerns about head injuries today may cause changes not just in the gear that players wear but also the rules of the game.

This welfare-based concern has also been used to justify bans on drugs that are unsafe.[48] The intake of anabolic steroids, for example, may increase the mass of the heart muscle, leading to a decrease in delivery of blood to the liver, raising the risks for both heart disease and liver failure. Taking excess EPO could increase the risk of blood clots and hence heart attacks and pulmonary embolisms. Gene therapy forms of doping can produce immune reactions that pose long- and short-term risks to athletes' health. Drugs may also be banned to prevent players from being coerced into using them by their coaches, or by government entities in state-sponsored games.

By understanding these welfare-based concerns, we can also see why some enhancers are permitted. Caffeine and creatine supplements, for example, are allowed because they are generally viewed as safe, even though they do provide some competitive edge.

Mind Games

We may want to prohibit the use of cognitive enhancement when zero-sum games are played with our minds. When the World Chess Federation sought to have chess classified as an Olympic sport, they adopted a "cognitive doping" prohibition for tournament players for all the drugs banned by the International Olympic Committee (IOC), many of which are stimulants with known beneficial effects on concentration, focus, and memory. When scientists studied over three thousand games played by forty chess players, they discovered that at least two drugs could substantially boost players' scores: modafinil, by an average of 15 percent, and Ritalin, by 13 percent. Even caffeine, the most commonly used cognitive enhancer, improved performance by an average of 9 percent![49]

This is why grandmaster Vassily Ivanchuk was found guilty of cognitive doping by the International Chess Federation after he refused to submit a urine sample for a drug test at the Chess Olympiad in Dresden in 2008. He was later exonerated, but only because he had not understood that the testing was mandatory.

Is it really cheating when chess players use cognitive-boosting drugs during tournaments? It is if the rules say that it is! The World Chess Federation and the International Chess Federation's reason for banning these drugs was to bring them into compliance with IOC regulations, to have a shot at being classified as an Olympic sport. That IOC ban is in place to preserve the "spirit of sport" and to protect athletes against the health risks that certain drugs may pose.

But just because a game is zero-sum doesn't mean that everyday competition should be treated similarly. Life is not a zero-sum game, and cognitive enhancement in everyday life stands to benefit everyone by lifting us up as a whole.

A Health Rationale in Life?

Laws, regulations, and norms have important purposes both in sports and in the larger society. Some are designed to nudge our behavior. Whether it's helmet laws for motorcycle drivers, seat belt laws for cars, food labeling laws to guide our purchasing behavior, or speed limits to limit the number of automobile accidents, our society is full of measures that promote general welfare through regulations, particularly when the societal costs of doing otherwise are considered too high. But when they are used instead to interfere with purely self-regulating choices, society begins to encroach on individual liberties without sufficient justification to warrant doing so.

We weigh societal interests against individual liberty interests all the time. The individual interest at stake in cognitive enhancement is the right to decide if, when, and how we want to improve our brains to enable us to navigate life with additional capabilities and benefits. When what is at stake is so fundamental to what it means to be human, the rationale for society's limitations of those choices must be quite convincing. If a cognitive enhancement led every person who used it to have uncontrollable fits of rage, we might rightly restrict access to that drug to protect others in society.

Athletics and chess are artificial, man-made systems. But life is not, and the right to self-determination over our brains and mental experiences in our everyday lives is much more important than any game.

Will the Pressure to Enhance Become Coercive?

One societal interest often advanced to justify banning or limiting the availability of cognitive enhancers is that their very existence creates both implicit and explicit pressure to join a "race to the top."

While we will explore the idea of explicit coercion in a later chapter on brain manipulation, let's explore the idea of implicit coercion now.

Making cognitive enhancers available to individuals gives them a choice—the choice to enhance or not to enhance. This choice, which as we'll explore in the next chapter is tantamount to our relative right to self-determination, is part of the "bundle of rights" that cognitive liberty protects. It's true that a world in which enhancers are freely available may lead to their more widespread use, and that the more widespread their use, the more likely it is that people who choose not to enhance will fall behind. But this weaker claim of coercion doesn't justify social limits on enhancement.[50] We can and often do freely choose *not* to do something that makes us more competitive, choosing to take a vacation rather than working longer hours, spending time with friends instead of studying more for an exam—or gambling and doing things that can potentially make us less competitive, like eating a tasty dessert instead of a virtuous salad while training for a marathon.

Moreover, the idea that peer pressure would create coercion to take cognitive enhancers isn't borne out by the evidence. Multiple studies have shown that individuals are no more willing to take cognitive enhancers when others do so, but they *are* more likely to avoid them when others disapprove of them.[51] It isn't the availability or even the use of cognitive enhancers by others per se that leads to coercive pressure. Rather, it's when others perform better that pressure mounts to improve one's own performance to remain competitive. Which, of course, can be said of many other endeavors.

Back in 1961, Kurt Vonnegut's short story "Harrison Bergeron" envisioned a world in which no one was allowed to have an advantage over anyone else, whether "natural" or artificially induced:

> The year was 2081, and everybody was finally equal. They weren't only equal before God and the law. They were equal every which way. Nobody was smarter than anybody else. Nobody was better

> looking than anybody else. Nobody was stronger or quicker than anybody else. All this equality was due to the 211th, 212th, and 213th Amendments to the Constitution, and to the unceasing vigilance of agents of the United States Handicapper General. So:
>
> . . . It was tragic, all right, but George and Hazel couldn't think about it very hard. Hazel had a perfectly average intelligence, which meant she couldn't think about anything except in short bursts. And George, while his intelligence was way above normal, had a little mental handicap radio in his ear. He was required by law to wear it at all times. It was tuned to a government transmitter. Every twenty seconds or so, the transmitter would send out some sharp noise to keep people like George from taking unfair advantage of their brains. . . .

To ban cognitive enhancements because of "implicit coercion" is to follow the same train of logic that leads to Vonnegut's dystopian egalitarianism. The role of government should not be to make all our capabilities equal, but to enable us to flourish as individuals and as societies.

And while some may argue that each new potential source of inequality is one that we ought to prevent, a different approach would be to remove the barriers to cognitive enhancements by making them cheaper and more widely available. It may be better to try to lift humanity together rather than keep everyone similarly behind.

Is There a Duty to Disclose Enhancements?

While it may be morally and politically unjustified to limit access to cognitive enhancers without significant countervailing societal interests, should society nevertheless require certain individuals to disclose their use of them? Or the discontinuation of their use?

Consider, for example, the case of a physician who took cognitive enhancers throughout high school, college, medical school, and as she sat for her qualifying exams and board certifications. Once she begins her medical practice, she decides to stop taking them. Does she have a duty to inform her patients that she achieved her status as a physician with their help, but that she no longer uses them? Or is what it takes to get you through training different from what makes you a well-qualified doctor? Are her privacy interests outweighed by society's interest in having an accurate representation of her capabilities? Should we apply a necessity and proportionality analysis to decide?

In the United States, few states require physicians to disclose specific information about themselves. But in some legal cases, courts have held that doctors have a legal duty to disclose their level of experience with a given technique when a reasonable person would expect to be told this information. In one case, a man named Alan Andersen underwent a Bentall heart procedure at the Iowa Heart Center. His physician, P. C. Khanna, did not have any experience or training in performing the procedure, and Andersen suffered serious side effects that ultimately caused him to need a heart transplant. Andersen sued, arguing that he had a right under informed consent laws to know the limits of Dr. Khanna's experience.

The district court that initially heard Andersen's claim found that a physician does not have a duty to disclose "physician-specific characteristics or experience in obtaining informed consent." But the Iowa Supreme Court disagreed, holding instead that if a reasonable person in the patient's position would consider the information material to their decision about whether to undergo a treatment, a failure to disclose that information could breach the physician's duty to that patient.[52]

Is prior use of cognitive enhancers material to whether a patient would choose to be treated by that physician? And if so, what other circumstances would raise the same duty of disclosure? Would a student who took cognitive enhancers when they sat for the SAT need to disclose that on their college applications? Must a lawyer, pharmacist, or other professional who used cognitive enhancers during their education disclose that to all their clients?[53]

While our right to self-determination favors our choice to use cognitive enhancers, it isn't an absolute human right like freedom of thought, which means society *can* impose limitations or duties upon us when it is necessary to do so for the greater good. Whether that is achieved through regulatory hurdles, to ensure that the drugs are safe and effective for use, or through duties of disclosure on individuals, the right to self-determination is a relative right that must be balanced against society's interests as well.

We, as a society, will have to decide when we have a right to demand disclosure, and when individuals have a right to keep that information to themselves.

And as we will see in the next chapter, societal interests may be even stronger when it comes to braking or diminishing our brains instead.

6

Braking the Brain

Beatriz Arguedas, a subway conductor in Boston, Massachusetts, was driving her normal route on the Red Line on September 30, 2005, when a man jumped off a platform as her train pulled into a station. "We sort of made eye contact," she recalls. "I felt the thud from him hitting the train and then I heard a cracking sound underneath the train. And then, of course, my head starts thumping."[1]

Traumatized, Beatriz sought help in the ER, where she encountered Dr. Roger Pitman, a psychiatrist at Harvard Medical School who studies post-traumatic stress disorder (PTSD). Neither Pitman nor anyone else could predict whether Beatriz would go on to develop PTSD. But if she did, even years later the smallest trigger—a sound, a smell—could bring back the full horror of what she had just experienced. Faced with that uncertainty, Pitman suggested she enroll in an experimental study for off-label use of propranolol (a drug that is usually prescribed for high blood pressure). If she took it immediately, he told her, her brain might not store all the emotional memories of what she had witnessed.[2]

The hippocampus functions like the RAM (random-access memory) of a computer. It processes and temporarily stores new memories for a short time before they are transferred to the cortex, where consolidation for long-term storage occurs. When traumatic events occur, our bodies release stress hormones like adrenaline, which enhance memory consolidation. This is why people remember traumatic experiences so vividly, despite being unable to remember what they had for breakfast the day before. This trait likely evolved over many generations; the advantage it confers is that those strong memories will help us to avoid similar situations in the future. But the disadvantage can be crippling anxiety, insomnia, irritability, and self-destructive behavior. Dr. Pitman's trials were designed to find out if propranolol could disrupt Beatriz's memory files before they were consolidated, reducing the likelihood that she would develop PTSD.

Surely, if the drug is safe, there is no moral or other reason Beatriz should not have the option of reducing her future suffering. But should the choice be hers alone? What if there are social costs to her doing so, such as making her an ineffective witness in a police investigation? Or if the side effects change her personality in ways that make her a danger to others? Does it matter whether the drug diminishes or enhances her brain functioning if it helps her cope later on?

Some of these questions were lurking in the recesses of my own mind when I took propranolol for the same reason. Our second child, Callista, had been hospitalized after contracting respiratory syncytial virus. As the days she spent in the this is the pediatric cardiac intensive care unit (PCICU) stretched into weeks, it became harder and harder for me to process the unfolding trauma. After a particularly harrowing night in the PCICU, I asked my neurologist to prescribe propranolol in the hope that it might help me endure, and perhaps lessen the likelihood of my developing PTSD, and he did. Our daughter did not recover; she died on Mother's Day in 2017.

While the early reports about propranolol were encouraging, it didn't work for me; I did go on to develop PTSD, suffering from vivid and visceral flashbacks of images from our first visit to the emergency room to Callista's gut-wrenching cries in the weeks that followed. I'm not alone in my failed response to propranolol. A recent meta-analysis suggests that it has no beneficial effects on PTSD.[3] Samuel Schacher, Emeritus Professor of Neuroscience at Columbia University, explains that "a given memory is very sparsely encoded. What that means is that our cerebral cortex, where most of these memories are stored, has about fifteen billion nerve cells, and a particular memory may involve a change of only a couple hundred of them. Finding those few hundred cells is very, very complicated."[4]

Neurotechnology may offer a better hope. In one technique, known as decoded neurofeedback (DecNef), an individual lies inside a functional magnetic resonance image scanner while recalling a traumatic memory so that machine learning algorithms can map the precise areas of the brain that it activates.[5] Once the neural mapping is complete, the person "erases" those memories through implicit neural feedback.[6] Just as I learned to make the birds chirp when I meditated with neurofeedback, people learn to achieve the target brain state by a process of trial and error—thinking of a scary movie, singing a song in their head, or doing a math calculation. Pretty quickly, they successfully induce it without being consciously aware of the trauma that it is associated with, and receive a reward, like a visual representation of a rising thermometer or hearing the sound of birds chirping. Through this process of repeated implicit memory reactivation paired with rewards, the individual retrains their brain.[7] Instead of revving the brain, the idea here is to brake it. I haven't tried it yet; should it become more widely available, I will.

We Allow Brain Braking All the Time

People brake their brains in all kinds of ways. Alcohol's brain-braking effects are well documented. It interferes with activities in the frontal lobes, limbic system, and cerebellum, leading individuals to feel disinhibited and act impulsively.[8] Many people who get drunk do so *for those very reasons.*[9] A college student at a party gets drunk purportedly to relax and have fun, a lawyer has a nightcap after a long day at the office, and my husband, Thede, and I join our monthly "wine club" as an excuse to connect with our friends in a social setting.

Many societies across the world permit the consumption of alcohol, with some important limitations. So long as you are of age, not driving, and not making a public scene, "being drunk" is entirely permissible in many cultures—perhaps even encouraged.

Of course, the United States and many other countries have tried to prevent individuals from consuming alcohol, and some societies and cultures still do so today. In 1919, the US Congress passed the Eighteenth Amendment, outlawing the manufacture, transportation, and sale of alcohol. In a piece written in support of Prohibition at the time, the Reverend Floyd W. Tomkins described "the menace of intoxicating drink to the peace and safety of a community."[10] The behaviors of drunken individuals have not changed since those days, but societies' willingness to accept such behaviors has; many states are now legalizing the recreational consumption of marijuana as well.

Does cognitive liberty include the freedom to choose to diminish one's brain and mental experiences, just as it includes the right to enhance them? The battle over motorcycle helmets sheds a stark light on the question.

An iconic fashion model who has graced the cover of *Vogue* more than twenty-five times, Lauren Hutton was the original supermodel. She is also an insatiable thrill-seeker who has wrestled alligators and rides motorcycles. In October of 2000, the then fifty-six-year-old Hutton participated in a hundred-mile celebrity motorcycle ride commemorating the opening of the Guggenheim Hermitage Museum and its inaugural exhibition *The Art of the Motorcycle.* Two and a half hours into it, she pulled over for a break as the heavy wind was making her eyes tear involuntarily. The English actor Jeremy Irons offered her his extra helmet, which had a visor, and insisted that she wear it. Hutton reluctantly put it on. That choice would save her life.[11]

Three minutes later, Hutton slid into a curve at a speed of more than ninety miles per hour. After skidding a hundred feet, she was launched into the air. A friend recalled the almost cartoonish scene they witnessed as they rounded the same curve: Hutton hanging twenty feet in the air, her legs sprawled out against the crystal clear desert sky. She came down hard on a hill of rocks breaking both her arms and legs, crushing three of her ribs, and puncturing a lung, and skidded another 170 feet facedown on her visor. She was in a coma for two and a half weeks. Though her initial prognosis was poor, less than a year later, she climbed aboard another bike to film a Tropicana commercial.[12] "I wouldn't have prefrontal lobes if I didn't have a helmet with a visor on," she later reflected.[13]

Each year, more than sixty-nine million people suffer from traumatic brain injuries (TBIs), the leading cause of death in motorcycle crashes.[14] TBI occurs after a sudden trauma, often a blow or jolt to the head, which causes damage to the brain. They range from a mild concussion, which may cause temporary confusion or headache, to severe ones leading to coma or death.[15] Simply by wearing a helmet, Hutton reduced her risk of death by 42 percent and her risk of a TBI by 69 percent.[16] But despite the overwhelming evidence that helmets save lives, only twenty US states, the District of Columbia, and Puerto Rico require all motorcyclists to wear them. Another twenty-seven states have helmet laws that apply only to minors, and three states (Colorado, Illinois, and Iowa) have no helmet laws at all.[17] The United States is an outlier in this regard. Forty-nine countries have comprehensive helmet laws in place, and most have at least some laws mandating helmet usage.[18] A little over half a century ago, forty-seven states, DC, and Puerto Rico all had mandatory motorcycle-helmet laws too.

Why has the United States moved *away* from mandating motorcycle helmets when the rest of the world has gone in the other direction? According to public health scholars Marian Jones and Ronald Bayer, the answer dates to the 1940s, when the market for motorcycles in the United States was shaped by returning veterans who had learned to ride military-issue Harley-Davidsons while overseas. Few of them wore helmets or even goggles, and only three states had helmet laws. In 1966, as public health rationales started to gain more support as a legitimate countervailing interest to individual liberties, the US Congress passed the National Highway Safety Act, which included a novel provision targeting motorcyclists—funding for future highway safety programs was made contingent on states passing helmet laws. In less than ten years, nearly every state had complied. California was the sole holdout, because it had a strong anti-helmet lobby.[19] The police power of states, the lobby successfully argued, did not extend to risks individuals *choose* to take.

The Supreme Court disagreed; in *Simon v. Sargent* (1972), it upheld a lower court decision that maintained that helmet laws were indeed in the public interest and not just paternalistic interference with the individual rights of motorcyclists.[20] Lower courts had been in near-uniform agreement that the laws were in the public interest. In one of those cases, a plaintiff had invoked John Stuart Mill to argue that the choice to wear a helmet was entirely personal. The court found otherwise, because "society picks the person up off the highway; delivers him to a municipal hospital and municipal doctors; provides him with unemployment compensation if, after recovery, he cannot replace his lost job, and, if the injury causes permanent disability, may assume the responsibility for his and his family's continued subsistence. We do not understand a state of mind that permits plaintiff to think that only he himself is concerned."[21]

Unconvinced, in 1975, motorcycle clubs, associations, and gangs descended on Washington, DC, for a mass protest, and the House Committee on Public Works and Transportation agreed to reconsider its stance. In May 1976, President Gerald Ford signed a bill that eliminated the helmet provision of the NHSA. During the four years that followed, twenty-eight states repealed their mandatory helmet laws, and deaths from motorcycle accidents increased by 20 percent. Some public health experts advocated for reinstatement of the laws, while others took the side of the motorcycle lobbies. Society permits individuals to take on all kinds of risks, they said, including participating in dangerous sports like rock climbing, where helmets are not required. That is the view that still prevails in the United States, thanks to what Jones and Bayer call the primacy of individual self-determination that Americans place over laws that protect "people from self-imposed injuries and avoidable harm"—an attitude that likely informed Americans' negative responses to mandatory mask and vaccine laws during the COVID-19 pandemic.[22]

Which side of the debate has it right? Does self-determination include the unfettered right to damage or destroy one's own brain? Do the costs of social services give society a say in choices that burden those systems? Or are we asking the wrong questions?

The Principle of Self-Determination

Self-determination is best understood as the right to be free from governmental interference when making decisions that only affect oneself. Often described as individual autonomy, the principle underlies many fundamental rights recognized across European, British, and US laws. But it is as misunderstood in the popular mind as it is fiercely protected.

The philosophical roots of self-determination trace back to the ethical theories of Immanuel Kant and John Stuart Mill, who described it as securing to individuals the capacity for free choice. But neither Kant nor Mill ever argued that the right to choose is limitless. Both recognized that it is circumscribed by the liberty interests of others.[23]

For Kant, who believed that rational agency and the capacity for self-determination set humans apart from other species, respect for individual autonomy was tantamount to being human. Kant and Mill both described the right to individual autonomy as a negative freedom (a freedom from external restraint) and, more broadly, the positive freedom to be in control of one's own life.[24] Mill extended the idea of autonomy to a broad conception of political liberty, as well.[25] Individuals, he wrote, are "the proper guardians of their own interests" and governments owe them nothing but the duty to prevent others from interfering with their liberty.[26]

Medical decisions are an example of purely self-regarding choices. Take my choice to use propranolol when our daughter Callista was hospitalized. Noninterference would mean that I should neither be forced to take the drug nor prevented from doing so, so long as it doesn't make me unable to perform my duties to others—most important, my duty as a medical decision-maker for our child.

Setting aside special cases like sports and mind games, cognitive enhancement is also a purely self-regarding choice that is fundamental to human flourishing. Government interference with it thus runs afoul of self-determination as a necessary element of cognitive liberty. The societal costs would have to be exceptionally high to justify interference with the choice to self-enhance. But does the same apply to the right to cognitive *diminishment*?

While international human rights law does not answer the question for us explicitly, a careful reading of the Universal Declaration of Human Rights (UDHR) suggests that an individual right to self-determination is a necessary precondition for all the individual rights it enumerates, including the right to be equal in dignity (Article 1 of the UDHR); to be free from discrimination (Article 7); to privacy (Article 12); freedom of expression (Article 19); and the right to one's own personality (Article 22, which secures to an "individual economic, social and cultural rights indispensable for his dignity and the free development of his personality").[27] In *Jehovah's Witnesses of Moscow v. Russian Federation,* the European Court of Human Rights (ECtHR) recognized the same, opining that "The very essence of the Convention is respect for human dignity and human freedom and the notions of self-determination and personal autonomy are important principles underlying the interpretation of its guarantees."[28]

The right to be free from discrimination protects against societal distinctions between the rights protected by the UDHR, which makes problematic societal distinctions between the choice to enhance or to diminish. Freedom against arbitrary interference with privacy has been interpreted by the European Convention on Human Rights to include the right to personal identity and a space to be in control of one's own faculties,[29] just as freedom of expression includes the freedom to develop one's own voice and identity. The right to privacy in Article 12 of the UDHR shields individuals against interference with their purely personal choices, without regard to whether we socially judge those choices as life-enhancing or diminishing. Just as the ECtHR recognized the individual right to self-determination, so too should international law be updated to explicitly recognize a right to self-determination as underlying the human rights guarantees of the UDHR.

These rights are circumscribed by Article 29, which incorporates a necessity and proportionality test on government restrictions on individual human rights: "In the exercise of his rights and freedoms, everyone shall be subject only to such limitations as are determined by law solely for the purpose of securing due recognition and respect for the rights and freedoms of others and of meeting the just requirements of morality, public order and the general welfare in a democratic society." Which means that just as Kant and Mill recognized a limit on free choice by individuals, so, too, does international human rights law. Self-regarding choices are those that do not affect the interests of anyone other than oneself.[30] But when a "person is led to violate a distinct and assignable obligation to any other person or persons, the case is taken out of the self-regarding class" and can be limited by law and public opinion.[31]

Each person in society exists relative to others, within a broader community and society of individuals who are also entitled to liberty.[32] Kant argued that self-determination also creates an obligation to not only avoid "intentionally withdrawing anything from the happiness of others, but also to try to further the ends of others."[33]

Mill believed that our obligations to others could limit the extent to which we could exercise our self-determination, explaining that "the mischief which a person does to himself, may seriously affect, both through their sympathies and their interests, those nearly connected with him, and in a minor degree, society at large. When, by conduct of this sort, a person is led to violate a distinct and assignable obligation to any other person or persons, the case is taken out of the self-regarding class, and becomes amenable to moral disapprobation in the proper sense of the term."[34] If our individual use of neurotechnology leaves us unable to care for a dependent child, for example, we are no longer acting in a self-regarding way, and society could rightly regulate our actions.

Self-determination over one's brains and mental experiences builds upon these views. It requires that an individual be free from governmental interference *unless their choices directly affect the interests of others.* This means that purely self-regarding choices over our brains—like tracking our own brain data and ordinary cases of cognitive enhancement or diminishment—should not be subject to societal interference. When our choices *do* interfere with the interests of others, any government interference with our choices must adhere to the principles of legality, necessity and proportionality.

It may very well turn out that diminishing our brains incurs greater societal costs than enhancing them does—by rendering us less able to fulfill our duties to others, by requiring others to scrape us off the pavement when we're injured or killed, and to care for us if we're unable to care for ourselves. And *that* is the right question to answer when we ask whether we have a right to diminish our own brains—whether the costs to society of our doing so are sufficiently high to justify societal intervention. The same argument is used to justify laws against drug abuse.

Breaking the Brain

The not-infrequent throbbing in my head began when I was ten or perhaps even younger. My well-intentioned parents attributed it to inflamed sinuses and treated me with eucalyptus steam baths and hugs. It wasn't until after college—when, by a stroke of good luck, my work at a strategy consulting firm brought me into contact with an insightful primary care physician whose office was in the same building—that I learned the truth. When I told him about my throbbing "sinus pain," he kindly explained that "those are migraine headaches." And the migraine medication he prescribed completely changed my life for the better. Along with it, he prescribed an opiate for acute migraine pain. When I handed my prescriptions to a pharmacist later that same day, I remember being perplexed by the suspicious look she gave me. It gave me the distinct feeling that I was doing something wrong.

Since then, the world has been turned upside down by opioid addiction. Starting in the early 2000s, overdoses and deaths from prescription opioids began to spike. This was partly attributable to OxyContin—a controlled-release version of the potent opioid drug oxycodone that I had been prescribed. Recreational use of OxyContin grew from four hundred thousand doses in 1999 to over 2.8 million by 2003.[35] By October 2017, the secretary of Health and Human Services had declared a public health emergency.[36]

Combatting the opioid epidemic has proved exceptionally difficult. More than five hundred thousand deaths a year arise from drug misuse, and more than 70 percent of those are related to opioids.[37] Over $740 billion is lost annually in reduced workplace productivity, health-care expenses, and crime-related costs.[38] Despite a multidecade, multibillion-dollar effort to combat the crisis, prescription painkillers remain easy to find, easy to use, and relatively cheap to buy on the black market. New drugs have been developed to help addicts kick their habits, and more addiction treatment centers have opened worldwide.[39] But the battle against opioid addiction is at a stalemate at best. More policing, arrests, and jail time are unlikely to reverse these tragic trends,[40] as these drugs take hold of the brain and don't let it go. The dopamine rush that they cause is so pleasurable that people will go to any length to experience and reexperience it.

Does the epidemic justify banning the sale of opioid drugs? Does the high risk of addiction justify limiting access to them, even for people who suffer from chronic pain? In 2016, the CDC issued guidelines that encouraged doctors to prescribe them less—which many doctors and insurance companies interpreted as guidance to avoid their use altogether. But not all patients who are prescribed opioids for chronic pain misuse them. In fact, only 8 to 12 percent of them become addicted. The result has been the systemic undertreatment of pain, with consequences that are every bit as devastating as addiction.[41] While I have been among the fortunate few whose chronic migraine pain is adequately treated, I am regularly made to jump through hoops to fill my prescriptions.

We may all agree that governments ought to try to mitigate the high societal costs of opioid addiction. But does doing so unduly interfere with the self-determination of individuals who have a self-regarding interest in controlling their own pain, brains, and mental experiences? How do we decide when the external costs to others sufficiently outweigh their own interests? How do we decide if, when, and how to intervene?

Almost every known addictive substance causes dopamine to be released in the nucleus accumbens in the brain.[42] Burning away a small section of this so-called pleasure center can stop the craving that defines addiction.[43] In 2003, Chinese scientists reported that they had successfully resectioned the nucleus accumbens of eleven patients. Within a year, another thousand underwent the surgery.[44] But international controversy followed as devastating side effects surfaced, including profound personality changes and memory loss. The Chinese Ministry of Health quickly responded by putting a moratorium on the procedure.[45]

Fortunately, the neurotechnological development known as deep brain stimulation (DBS) offers a more targeted intervention for willing participants. Gerod Buckhalter was one of the first to undergo this procedure. Once described as "Mr. Everything" by his local newspaper—he'd been a high school basketball and football star—Buckhalter fell far from grace at age fifteen, when he suffered a shoulder injury and was prescribed a six-week regimen of opioids to treat the pain. That became his gateway to heroin and other drugs. When offered the opportunity to undergo experimental DBS surgery at age thirty, he jumped at the chance.[46] A neurosurgeon drilled a hole in his skull and inserted an electrical probe in his brain. Then he was shown images that induced his craving for drugs while the probe delivered electrical signals that effectively reprogrammed the neural circuits underlying his reward pathways.[47]

Compared to ablative brain surgery, DBS holds greater appeal for the individual and society because it has fewer side effects, greater precision, and is reversible.[48] Other techniques using neurotechnology involve the use of transcranial magnetic stimulation from a device held outside the head that targets ultrasound in high- or low-frequency waves to reach structures deep inside the brain.[49]

The nucleus accumbens is just one of the parts of the brain that these techniques target, and addiction is just one of many behaviors that they treat. Much as the nucleus accumbens triggers pleasure in response to opioids, the amygdala triggers fear.[50] An amygdalotomy—the surgical removal of the amygdala—could, in theory, prevent an individual from having panic attacks. But ablating the amygdala can also cause profound personality changes, blunting emotions to such an extent that the person becomes a psychopath. Nearly all serial killers show psychopathic tendencies, such as a lack of remorse and apparent disregard for the loss of human life.[51] Some of their behavior can be explained by the lack of a fear response in their brains.[52]

Does the risk of side effects to the individual and to society outweigh the potential benefits of these interventions? Do the external costs of the behaviors they are meant to change justify those risks? "External costs" means that a person doesn't bear all the costs of their choices. Take cigarette smoking. An individual smoker may believe that cigarette smoking is purely self-regarding behavior. But studies estimate its external costs at $323 billion every year. That said, less than $5 billion is imposed outside smokers' own families, or less than 2 percent.[53] Do *any* external costs justify limiting or banning smoking?

Estimates of the external costs of risky behavior depend on the social context in which an individual makes their choices, including the availability of health insurance and health-care structures, medical practices and technologies, and scientific knowledge.[54] Alongside those empirical factors are many subjective ones that our biases can lead us to over- or understate.[55] The Reverend Floyd W. Tomkins's assessment of the "menace to the peace and safety of a community" that alcohol poses might not be the same as our own.

Certain choices may be more likely to have external costs than others. The choice to diminish our brains with medications that block pain receptors while releasing a rush of dopamine has high external costs because it renders any given individual more likely to become addicted, limiting the scope of their future autonomous choices. My choice to erase my memory of an event I have witnessed might also have more easily measurable societal costs than my choice to enhance my memory.

But Mill cautioned us against adding too many remote externalities into our calculations. While we are all interrelated, is it fair to characterize all these costs as *direct* costs that limit the liberty interests of any others? The mere fact that a negative externality exists does not always justify interfering with people's freedom of choice. Moreover, governmental interference with individual choices is unlikely to affect all people equally. A simple cost–benefit analysis can elide issues of equity.[56] The racial disparities in drug arrests have been well documented. Restrictive regulations on pain medications, for example, are more likely to harm poor people than rich, who can shop for more sympathetic doctors.

What all of this means is that when an individual chooses to brake their brain for a self-regarding reason—such as to break an addiction—we ought to tread lightly before we intervene. And we ought to clearly articulate ex ante what societal harms we will consider when making the decision about when to do so.

The Right to Self-Determination over Our Brains and Mental Experiences

Some brain hacking impose costs on others that, depending upon their nature and degree, may justify some interventions. When we drink alcohol, we have slower reaction times, so society prohibits us from driving a motor vehicle. But most Western societies do not outright prohibit individuals from drinking. This may be because the impacts of drinking on most individuals—and society—are temporary and transitory, rather than permanent and unavoidable. We may weigh other interventions that have more permanent effects differently. But as for the broader question, whether individuals have the right to enhance or diminish their own brains, the answer is that they do—so long as the externalities of their decision does not unduly interfere with the liberty interests of others.

Like other rights in the bundle of sticks that make up cognitive liberty, the right to self-determination over our brains and mental experiences requires us to explicitly update our understanding of international human rights law. Self-determination is not an absolute right—rather, it is a liberty interest that is limited only by the liberty interests of others. This understanding undergirds our right to be equal in dignity, free from discrimination, to have our privacy respected, and be able to express ourselves freely.

As Mill argued, "With regard to the merely contingent, or, as it may be called, constructive injury which a person causes to society, by conduct which neither violates any specific duty to the public, nor occasions perceptible hurt to any assignable individuals except himself; the inconvenience is one which society can afford to bear, for the sake of the greater good of human freedom."[57] Put simply, contingent injuries don't rise to satisfy the necessity test for government interference. When society interferes with purely personal conduct, it invariably does so wrongly and in the wrong place. We can and should avoid acting as the moral police by focusing our inquiries strictly on whether our choices harm other people. If they do not, we ought not interfere with them, as they are purely self-regarding ones.[58]

This does not leave society powerless to influence our choices over our brains and mental experiences. We should provide people with information about the risks and benefits of certain brain interventions, offer incentives for choices that align with the common good, and even consider the necessity of restricting certain behaviors—perhaps driving a motorcycle without a helmet—to limit the financial burden that an individual's choices can impose on society.

But here, too, we should tread lightly. John Stuart Mill cautioned that while the state has an interest in discouraging injurious conduct like drunkenness, to "tax stimulants for the sole purpose of making them more difficult to be obtained, is a measure differing only in degree from their entire prohibition; and would be justifiable only if that were justifiable . . . [citizens'] choice[s] of pleasures, and their mode of expending their income, after satisfying their legal and moral obligations to the State and to the individuals, are their own concern, and must rest with their own judgement."[59]

There is an inextricable link between self-determination and individuality. "It is not by wearing down into uniformity all that is individual in themselves, but by cultivating it and calling it forth, within the limits imposed by the rights and interests of others, that human beings become a noble and beautiful object of contemplation," wrote Mill.[60] We should not discourage individual choices simply because we wish for greater uniformity, or have particular views about how individuals ought to exercise their right to cognitive liberty. Respecting people's right to self-determination—including their right to enhance or diminish their brains—will further enable human flourishing. But that flourishing will be threatened if we don't better define what others can be allowed to do to our brains.

7

Mental Manipulation

Thanks to the five-year age gap between our youngest and oldest daughters, my husband and I have been witnessing anew the joys of discovering a world beyond one's own mind. When our youngest daughter, Alectra, first pointed her chubby index finger at one of her unicorn figurines and asked me, "Is that *your* favorite, too, Mommy?" I smiled as I showed her the pink one I preferred, and silently cheered that she had reached a new milestone, the recognition that her own preferences may be different from mine. As she started to distinguish between happy and sad faces in her illustrated *My First 100 Words* book, I celebrated her newfound power of mind reading.

No, she isn't a telepath, as far as I'm aware. Alectra was just making inferences about the mental states of others. This important stage in cognitive development is when we become aware that others have different beliefs, intuitions, plans, desires, and intentions than our own. The ability to infer what other people are thinking and to predict what they will do as a result of their internal mental states is known as theory of mind in psychology.[1] It's called a "theory" because it's exactly that. Without actual or direct knowledge about what another person is thinking, we make inferences.[2] By observing the nonverbal clues of their biases, beliefs, and desires, we can not only develop a deeper understanding of their thoughts than what they may choose to share with us verbally but also predict their likely behaviors. Once Alectra discovered that I am afraid of wasps, for example, she could predict that I will dart away whenever one flies near me.[3]

Equipped with a theory of mind, we try to influence and persuade people around us,[4] and use the same tools to infer when others are trying to persuade us. Before a child develops a theory of mind, they can't recognize when someone is trying to influence them, such as when an advertiser tries to hook them on a new toy. But when they are as young as three years old, children begin to show early signs of understanding other people's intentions, and by age five they can usually understand their emotions and desires. By late preschool, they can even start to recognize embedded mental states: *Mommy is scared of wasps* (a first-order unidimensional desire), or *Daddy* thinks *that Mommy is scared of wasps* (a second-order embedded mental state).[5]

Children test out their newfound abilities at home. Alectra quickly pegged me as more likely to bend to her will than her father, so when she recently wanted a popsicle before dinner, she came to me with her request. "Mommy," she asked innocently, "would *I* like a popsicle?"

"I don't know, Alectra, *would* you like a popsicle?" I asked.

"Oh yes, thank you, Mommy!" she exclaimed as she reached for the freezer door. Somehow, I'd been hoodwinked again.

While everyday mind reading and persuasion are fundamental to what it means to be human, most people feel differently about being manipulated or coerced. We would all agree that it is wrong to hold a gun to someone's head to make them do or believe something. Or that Lord Voldemort's use of legilimency to penetrate the minds of his victims and bend them to his will in J. K. Rowling's Harry Potter series was evil.[6] But how do we know when we have left ordinary human persuasion behind and crossed the line into the morally fraught territory of coercion or manipulation? Do advances based on neurotechnology blur that line?

Marketing to Our Brains

Are you one of the millions of people who watched the epic science fiction film *Avatar* in 2009? I was utterly transfixed by the fictional world of Pandora—the moon that humans were colonizing, threatening its indigenous Na'vi. Its premise, that a human could remotely operate a genetically engineered Na'vi body with his brain to infiltrate their community and gain their trust caught me hook, line, and sinker.

Years later, I discovered that *Avatar*'s director, James Cameron, might have hooked me in other ways, too, by using what was then the cutting-edge technique of "neuromarketing"—a hybrid between neuroscience and traditional consumer research. When Cameron was filming *Avatar,* he told *Variety* magazine that he believed its 3D aspects would be particularly appealing to audiences because a "functional-MRI study of brain activity would show that more neurons are actively engaged in processing a 3-D movie than the same film seen in 2-D."[7] That spurred the firm MindSign to reach out to Cameron and offer him a freebie on their neuromarketing services. Cameron agreed, and MindSign took fMRI brain scans of research participants while they watched potential *Avatar* trailers to pinpoint which ones—and which parts of them—most engaged viewers.[8]

Moran Cerf, a neuroscientist and business professor at Northwestern University (and a leading researcher in neuromarketing for cinema) explains that good movie trailers "make all brains look alike. A good filmmaker can create an experience that will take over your brain. It is so powerful that it is actually working the same way on every other person that watches the movie."[9] A good film should make everyone's brain watching it show the same patterns of attention, immersion, and engagement. Who knows if it was the stunning

3D graphics, the acting, directing, the neuromarketing, or some other qualia that is as yet unknown, but *Avatar* was nominated for nine Academy Awards, including Best Picture and Best Director, won Best Art Direction, Best Cinematography, and Best Visual Effects, and became the highest-grossing movie of all time.[10]

Since then, neuromarketing has gone mainstream. Neuromarketing means using physiological and brain measurements to understand consumer motivations, preferences, and decision-making to inform marketing, pricing, and product development decisions. Companies leverage the tools of neuromarketing to make their branding and products more compelling, manipulating our unconscious minds to motivate our purchasing decisions.[11] The field started to gain steam in the early 2000s, when researchers began to publish novel insights about consumer preferences using its tools.

Hundreds of billions of dollars are spent annually on advertising.[12] But the effectiveness of those investments has long been stymied by people's lack of conscious awareness of the emotional, attentional and sensorial processes that underlie their choices, meaning they cannot reliably report on their reactions to products, advertisements, and brands.[13] Brain data promises to reduce the uncertainty of traditional marketing surveys by directly decoding consumers' unconscious preferences and biases.[14] Once just a pipe dream, neuromarketing has shown the power to predict consumer behavior in a series of studies.[15]

One of the first groundbreaking results came in 2004, when researchers tried to understand why some people have a strict brand allegiance to Coke or Pepsi, despite their near-identical chemical compositions. In blind taste tests, people struggle to recognize their "preferred" brand. But in unblinded taste tests, they consistently rate the brand they claim to prefer more highly. Rather than directly asking participants why, these researchers turned to their brains for answers. Participants were served Coca-Cola and Pepsi under blinded and unblinded conditions, while their brains were scanned using fMRI.

In blind taste tests, when participants drank Coca-Cola or Pepsi, they had greater brain activity in a region of the brain called the ventromedial prefrontal cortex, believed to integrate sensory, affective, and memory-related information. In the unblinded tests, people with a stated preference for Coca-Cola showed a sharp increase in brain activity in the brain regions associated with memory and cognitive control—their dorsolateral prefrontal cortex, hippocampus, and midbrain. Pepsi drinkers' brain activity looked the same in both blinded and unblinded tests. The difference between the two groups revealed something important—Coca-Cola's branding made those with a stated preference for Coke think and remember, and not just taste and feel when they drank a branded Coke. Pepsi drinkers weren't having the same experience. Coca-Cola (but not Pepsi) had managed to shape user preferences in a way unrelated to the taste and smell of the beverage. This "brand effect"—an association between the brand and how it made people think and remember—changed their subjective experience of pleasure and the decision-making that followed.[16]

A few years later, researchers at the California Institute of Technology found a similar effect in a study involving different priced wines. The researchers served research participants wine while scanning their brains with fMRI and told them (made-up) prices for each variety. Unbeknownst to the participants, all the wines had identical retail values. The participants nevertheless consistently preferred the more "expensive wines." Their brain activity showed a similar effect to the Coca-Cola branding. Changing the price didn't change their sensory experience of tasting the wine, but it did increase brain activity relevant to their subjective experience of pleasure during experiential tasks.[17] Armed with this information, wine producers can manipulate the price of wine without changing how it tastes and expect consumers to rate it more highly.

With insights like these, companies started to see neuromarketing as the "holy grail" of marketing—a reliable way to decode consumers' once-inaccessible unconscious reactions to packaging and pricing and measure their actual emotional engagements with advertisements. Paying attention to an advertisement meant it had cut through other distractions. Some studies even claimed that certain brain wave patterns (such as prefrontal asymmetry in the gamma frequency band, which you can detect with an EEG device) may be tightly correlated to our "willingness to pay" for a particular product.[18]

Not everyone was convinced. In 2017, Ming Su, a marketing professor at UC Berkeley, wrote in the *California Management Review* that "neuroscience either tells me what I already know, or it tells me something new that I don't care about."[19] Most businesses have long understood that higher-priced wines are perceived as more valuable by consumers, even when they taste the same. Su was far from the only dissenting voice. Academics were generally pessimistic in the early 2000s. Their pessimism was buttressed by early setbacks in the field, including inappropriate assumptions about the localization of particular psychological processes within the brain, and a lot of oversimplification about how we make decisions. The technology and software powering it also limited neuromarketers' ability to measure brain activity and analyze it accurately, and the field was stymied by methodological problems, including reverse inference, whereby researchers assumed a pattern of activation meant a certain cognitive process was engaged.[20]

But as the science, technology, and methodology supporting neuromarketing have progressed, brain data now offers real advantages when predicting the future success of consumer products. More portable fNIRS are being used to map areas of the brain engaged with ongoing psychological processes. EEG has become a popular mainstay in neuromarketing because of its ability to detect unconscious preferences and biases, despite some of its limitations, including its inability to pinpoint where in the brain activity occurs (including its inability to reach the deep, subcortical regions of the brain, where a lot of consumer decision-making occurs).[21] Physiological measures—heart rate, electrodermal activity, eye tracking, facial electromyography, and the like—complement these tools, enabling marketers to detect preferences, desires, and biases outside of individuals' awareness and control.[22] By 2017, the Advertising Research Foundation, a nonprofit industry association for creating and sharing knowledge in advertising, heralded neuroscience as better able to predict what consumers actually want and do.[23]

Today there is a burgeoning industry of more than 150 neuromarketing firms worldwide, ranging from technology providers to those that offer more comprehensive services.[24] Some companies, like NBC and Warner Bros. Discovery, have operated their own neuromarketing units for years, while others, like Microsoft, Google, and Meta, have formed units to do the same. Most of these companies believe that decoding the subconscious consumer mind is critical to their ongoing success.[25] It seems to be working.

Anyone who watches TV sees a lot of PSAs promoting worthy causes. You may even have responded to some of them with donations, to help families affected by wildfires, or those impacted by the tragic loss of life in a mass shooting in the United States. You may have donated to support Ukrainian refugees, what with more than two-thirds of the country's children displaced by the Russian invasion.[26] The United Nations has called it the "fastest and largest displacement of people in Europe since World War II."[27] But maybe not, if the call to action included images of their displacement's devastating effects. Why is that?

The *New York Times* reporter Charles Duhigg puzzled over a similar issue—why people weren't donating to Syrian refugee relief. One answer came from his interviews with the social scientists Jennifer van Heerde-Hudson and David Hudson, who have spent years studying how charities solicit donations. "Children who have lost their homes, starving families, the heartstring things," David Hudson told him. "That's what everyone believes works." But they found the opposite to be true. When campaigns shift from images of poverty-stricken children and messages like "Please donate before it's too late" to hopeful and inspiring images of children holding signs like FUTURE DOCTOR, people are more likely to give. "If you can trigger a sense of hope, donations go up," explained Mr. Hudson.[28] Or as Duhigg puts it, "It's not entirely your fault" if you aren't donating to refugees. "You just haven't been manipulated properly."

When neuromarketers tweaked an unsuccessful campaign by the Italian UNCHR for refugees, its new commercial led to a 237 percent increase in sellable calls over the prior one. The brains of test subjects showed them how to do it. The first commercial had low emotional arousal throughout, and poor engagement during the final call to action. Using EEG insights from participants watching the commercial, they modified the new commercial with new images to evoke greater empathy in viewers, and with new visual effects in the call to action that better engaged viewers' brains.[29]

Is Duhigg right to call this "manipulation"? The Advertising Research Foundation was concerned enough that it called for government standards to guide neuromarketing research.[30] The Neuromarketing Science and Business Association—established in 2012 to support researchers and practitioners of neuromarketing worldwide—developed a code of ethics to guide responsible progress in the field.[31] Neuroethicists have long called for protections for people who might be harmed or exploited by neuromarketing as its effectiveness improves, potentially eroding viewers' autonomy.[32] But there is little consensus on what constitutes permissible and impermissible uses of neuromarketing, despite growing anxiety about whether researchers might discover aspects of the human brain that could "turn individuals into buying robots and produce dangerous behaviors," like addiction or overconsumption.[33]

As giant corporations like Coca-Cola, McDonald's, and Procter & Gamble seek to understand consumption patterns by vulnerable populations such as children, addicts, and gamblers, many worry this research will be used to harm humanity. Anxiety over neuromarketing has been further amplified by its opaque use by tech giants to shape customers' experiences. Meta has already come under fire for running at least one psychological experiment without user consent—manipulating the moods of more than seven hundred thousand users by altering its newsfeeds with happy and sad content, just to see the results.[34]

Can a trailer, movie, or product be so enticing that we can't reasonably resist it? Does it matter if the tactic is put to uses that are intended to benefit versus harm us? Does it matter if the influence is hidden from our view?

Persuading or Addicting the Brain?

In 2017, the entrepreneur, venture capitalist, and former Facebook engineer Justin Rosenstein joined several other former Facebook executives to sound an alarm about the techniques social media companies deploy to target users' unconscious decision-making.[35] Rosenstein helped create the Like button (originally called the Awesome button) during his days at Facebook, which later became a standard feature across most social media platforms. It was intended, he explained, to "send little bits of positivity" across the platform.[36] But Rosenstein had since come to believe that it was profoundly harming humanity by addicting people to the platform and tying their self-esteem to the Likes they receive.

Rosenstein has joined the ranks of other leading technologists who are weaning themselves off products they helped to create, while sending their children to schools where iPhone, iPads, and even laptops are prohibited. I can't blame them. There is something profoundly disconcerting about having your barely verbal two-year-old wake up and scream, "I need your phone!" as her first words of the morning.

By 2018, the average Generation Z smartphone user was so enamored with their phones that they unlocked them at least 79 times a day.[37] Most of us are using our phones at least 20 percent more often than we did in 2015,[38] while 60 percent of college students believe they have become an addiction, and 87 percent of millennials admit that their smartphones never leave their side.[39]

There is growing concern about the societal implications of

this addiction, not the least of which is the interference with our ability to focus.[40] "Everyone is distracted," Rosenstein said. "All of the time."[41] Emory law professor Matthew Lawrence recently took on these issues by arguing for a right to freedom from addiction, citing the dangers of social media addiction and concern that "social media and game companies have knowingly used technology to plant repetitive, unwanted thoughts in users' minds, without their knowledge and consent; indeed, without even the basic 'warning: this product is addictive' that now appears on cigarette packages."[42] With harms ranging from increased risks of suicide to automobile accidents from distracted drivers and loss of workplace productivity, Lawrence argues that the US Constitution should be interpreted to address "the liberty implications of addictive technology" to bolster the case for legal interventions.[43]

But just as technologists like Rosenstein are abandoning ship, other technologists are paying high-ticket entry fees to attend conferences curated by Nir Eyal, the author of *Hooked: How to Build Habit-Forming Products,* to learn how to better addict people to their products.[44]

The everyday technologies we use "have turned into compulsions, if not full-fledged addictions," Eyal writes. This is far from accidental, but "just as their designers intended."[45] His website advertises "how to build habit-forming products" and invites companies to "Discover the secrets the world's leading technology companies use to keep users coming back—and apply them *right now* to your product."[46] Facing public scrutiny, by 2017 he started to acknowledge the growing anxiety about whether these were manipulative tactics, cautioning his audiences that they should be careful not to use them for harm. But he still stalwartly defends what he teaches, arguing that just as "we shouldn't blame the baker for making such delicious treats, we can't blame tech makers for making their products so good we want to use them."[47]

Tristan Harris, a former Google employee turned tech critic, cautions that we are "jacked into this system . . . All of our minds can be hijacked. Our choices are not as free as we think they are."[48] Like Eyal, Harris studied under B. J. Fogg, a behavioral psychologist at Stanford University who is well known for his mastery of the ways that technological design can be used to persuade. Their similarities end there. Instead of teaching Google how to exploit our brains, Harris cautioned them against doing so. As a result, he became their in-house design ethicist and product philosopher. I "got to sit in a corner and think and read and understand," he says. That's when he came to believe that LinkedIn exploits our need for social reciprocity to widen its network, how media platforms like Netflix use autoplay to keep our attention fixed as the service transitions from one episode to the next, and how Snapchat created Snapstreaks to keep communication between at near constant levels between its users.[49]

Each of these approaches exploits shortcuts in our brains. To keep ourselves safe, we pay more attention to fearful, dangerous stimuli,[50] so social media uses notifications and alerts to make our brains believe that we urgently need to turn our attention back to their platforms.[51] Children with a particular genotype are more likely to be addicted to nicotine if they vape their first ecigarette before they are fifteen years old;[52] it's perhaps knowledge of this which led tobacco companies to use flavored tobacco products and advertisements to target young people to addict them for life. Features like infinite scroll and algorithmic recommendations prey on evolved mechanisms in our unconscious without us even realizing they are doing so.[53] The algorithms have become so sophisticated that tech companies can even target specific people to hook them on their platforms.[54] "Once you know how to push people's buttons, you can play them like a piano," Harris declares.[55]

Companies have been doing some variation of this all along. Food companies have long manipulated combinations of salt, sugar, and fat to serve us foods that our bodies end up craving.[56] Politicians have long tailored their manner of dress and speech to make us believe they are people like us and hence trustworthy, to garner our votes at the polls. I remember a course I took to prepare for the

MCAT, where they counseled us to write in cursive rather than print for the essay portion of the exam, because cursive writers statistically scored higher. Have persuasion technologies just gotten better at giving us more of what our brains have wanted all along?

How, if at all, is this kind of persuasion different from Alectra turning the tables on me to make it seem like it was *my* idea to offer her a popsicle before dinner? Why do we smile when a young child gets her way but shudder when tech giants do the same? Is it because the child's manipulation is transparent while theirs is harder to detect and resist?[57]

Taking Advantage of Brain Heuristics

For years, I have studied when and why criminal defendants use brain scans or neuroscientific experts to argue that their "brain made them" commit a criminal offense. If you believe that all our actions, perceptions, and beliefs ultimately come from our brains, then describing a defendant's brain in greater details shouldn't have that much bearing on their culpability. But defenses that describe criminal conduct as arising from the brain rather than a bad choice regularly throw judges and juries for a loop.

Some of our credulity may be explained by what has been called the "seductive allure of neuroscience,"[58] a phenomenon in which people are swayed to think more favorably about psychological explanations if they include references to or images of the brain.[59] Even if the references are logically irrelevant or make a good explanation worse, people still rate arguments that include neuroscience as more persuasive. Do people just favor longer over shorter explanations? Or explanations that seem more authoritative? Perhaps scientific-sounding jargon is more inherently persuasive?

Dr. Deena Weisberg, a psychologist at the University of Pennsylvania, set out to find out. In a series of three experiments, she tested each of these hypotheses. She recruited participants to take online surveys in which they were shown good or bad versions of explanations. Each example included a description of a psychological phenomenon (such as that babies have the ability to do simple math; that there are gender differences in spatial reasoning; and that there are differences in seeing and imagining objects) and gave eight different explanations for each. The explanations were presented in two lengths. Some included neuroscience, and some made no reference to science at all. Each trial participant saw one of the phenomena, received one explanation, and was asked to rate that explanation on a seven-point scale ranging from very unsatisfying (−3) to very satisfying (+3). Subjects rated longer explanations as better than shorter ones, but neuroscience had an independent and stronger effect. While people could judge between good and bad explanations, adding neuroscience interfered with their ability to do so.[60]

Researchers have replicated these findings across other reducible scientific disciplines and found similar effects on our ability to tell a good argument from a bad one. To understand the significance of reducibility, "consider the relationship between chemistry and physics," explained Weisberg. "While it is logically possible that atoms (and other elements of a physical ontology) could exist without there being any molecules, it is not logically possible that molecules could exist without atoms. Atoms, then, are logically prior to molecules. If an explanation of a molecular phenomenon is translated in terms of atoms, and if the atomic translation does not omit any aspect of the molecular version of the explanation, then we can say that the explanation has been *reduced* from the chemical to the physical level."[61] When we are presented arguments with reductive reasoning, even incorrect reasoning, it makes it harder for our brains to pick out a good argument from a bad one.

While no one has yet established exactly why that may be, reductive explanations may misguide us into failing to think—using shortcuts our brains provide us instead, that make us more likely to believe the information before us without thinking critically about it.[62]

We use all kinds of cognitive shortcuts to help us navigate our environments more efficiently, and reductive scientific logic appeals to a hardwired heuristic that allows us to categorize reductive-sounding logic as correct.[63] Reductive logic may lead us to accept an argument quickly rather than carefully, and "people who believe false things . . . don't think carefully," claims David Rand, professor of Management Science and Brain and Cognitive Sciences at MIT.[64]

Using reductive science and other cognitive shortcuts as heuristics for judgments about argument quality makes us more susceptible to fake news and other misinformation. People spread misinformation intentionally and unintentionally. Those who do so intentionally often study how to frame their claims in ways that capture our attention and make us more likely to pass that information on to others—with sometimes devastating effects.[65] Misinformation has tanked stock prices, wiping out, in one case, $130 billion in stock value after a false tweet about Barack Obama being injured in a White House explosion went viral in 2013.[66] Misinformation about terrorist attacks or natural disasters can set off panics, undermining entire societies.[67]

Splashy headlines are another strategy that appeals to our unconscious cognitive biases. The more outlandish the claim, the more likely our brains are to focus on the novel stimulus.[68] Unexpected information can pierce our attentional filters by triggering our sensory cortex to pay attention, triggering the brain to release dopamine, which makes the attention that much more rewarding.[69]

These strategies to exploit our unconscious biases may partly explain how false information spread so widely during the COVID-19 pandemic. At the height of the pandemic, the not-for-profit Center for Countering Digital Hate found that anti-vaccine activists had reached more than fifty-nine million followers with messages that were shared over 812,000 times on platforms including Facebook, YouTube, Instagram, and Twitter. More than 65 percent of that anti-vaccine content was attributable to a group they dubbed "the Disinformation Dozen,"[70] many of whom have been spreading spurious medical claims for years.[71] Some of the most prominent members used A/B testing to see which of their stories were most likely to go viral. If we were to study the most successful of those claims, they likely preyed on unconscious neural processes, causing real and devasting harm to individuals and society. Exposure to even a small amount of misinformation reduced the number of people willing to take a COVID vaccine by up to 8.8 percent.[72]

We might believe we are impervious to tactics designed to make us susceptible to misinformation. But even the most well informed of us fall prey to them. The more we are exposed to the same false claims, the more our brains begin to mislead us into believing them. Becoming more deliberate and critical about what we read and hear can help us counteract those effects, making us less likely to contribute to the problem.[73]

What Mental Manipulation Is Impermissible—and When?

We are constantly bending and being bent to the will of others. And neurotechnology may be enabling newfound ways for those seeking to bend others to their will.[74] In chapter 3, I discussed the report by Dr. Ahmed Shaheed, the special rapporteur to the UN General Assembly on freedom of religion or belief, and his recommendation to expand the international human right of freedom of thought to include the right not to reveal one's thoughts nor to be penalized for them. He also recommended that freedom of thought include the right not to have our thoughts manipulated.[75] But manipulation is a slippery concept. If ill-defined, an absolute prohibition on it could do more harm to human interactions than good.

About a decade ago, I went down a deep rabbit hole when I was trying to untangle claims about philosophical and legal free will. The written debate goes back at least two thousand years, but neuroscientists have recently joined the fray by arguing that our decision-making is hardwired in our brains.[76] Punishment, they argue, cannot be justified by retributivism—an eye for an eye—because people are not morally culpable for their actions. I disagree and have sought in my own scholarship to explain why freedom of action is a freedom worth defending.[77]

In a well-known 1971 essay titled "Freedom of Will and the Concept of a Person," the American philosopher Harry Frankfurt describes what he calls a peculiar characteristic of humans—that we can form "second-order desires." Besides our subconscious preferences, biases, and desires, we can also "want to have (or not to have) certain desires and motives."[78] Frankfurt calls this capacity for reflective self-evaluation of those biases and desires higher-order volition. We don't have to be fully *aware* of our unconscious desires to engage in reflective self-evaluation. We might be completely unaware of some desires, while being mistaken about others. Free will, he argues, is our capacity to form higher-order volitions, by recognizing certain desires as our own.

Frankfurt uses an example of two animals addicted to drugs. One is conflicted about his addiction—he craves the drug but also wants to be free from it. He wants his desire to be free from his addiction to become the one that drives his behavior. The other animal also has conflicting desires but lacks the capacity for self-reflection, and so doesn't form a preference between them. The first animal is human while the latter is not, because only the first makes one of his desires "more truly his own, and in so doing, he withdraws himself from the other." Frankfurt implicitly connects this to manipulation, by explaining that when the human addict is unable to break his addiction, he feels like the force "moving him to take the drug is a force other than his own."[79] When we believe that something *other* than our free will is driving us to act contrary to a desire we identify with, we feel as if we are being manipulated.

Frankfurt's example helps us distinguish between freedom of will and freedom of action. Freedom of will is our capacity to identify with our desires. Freedom of action enables us to make our will our own through our actions. Our freedom of will may be illusory—we commit to desires, biases, or preferences believing we have done so freely, but we may have chosen that preference because it was unconsciously primed by our environment. Our freedom may also be interfered with, making it harder to make our volition effective, if we are manipulated into acting compulsively with a "force other than [our] own."[80] We may want to stop checking Instagram every five minutes, but cleverly timed notifications compulsively draw us back in.

In *Autonomy and Behavior Control*, Gerald Dworkin characterized a person's motivation as belonging to a person without it truly being "their" motivation, if that motivation is brought about by interfering with their ability to reflect rationally on their interests through deception, or by short-circuiting their desires and beliefs, making them a passive recipient of the change.[81] Philosophers Daniel Susser, Beate Roessler, and Helen Nissenbaum in a recent article defined manipulation in the digital age,[82] arguing that permissible influence appeals to our "capacity for conscious deliberation and choice," while manipulation takes "hold of the controls," depriving us of "authorship over [our] actions" and driving us "toward the manipulator's ends."[83]

Other scholars describe manipulation as interfering with our "mental integrity," which Andrea Lavazza describes as "the individual's mastery of his mental states and his brain data." He argues that we should draw a bright line that prohibits unconsented-to interferences that "can read, spread, or alter such states and data in order to condition the individual in any way."[84] Marcello Ienca and Roberto Adorno are more tempered in their claims, arguing for "specific normative protection from potential neurotechnology-enabled interventions involving the unauthorized alteration of a person's neural computation and potentially resulting in direct harm to the victim."[85]

These accounts all coalesce around a definition of manipulation as hidden attempts to use our cognitive biases, emotions, or subconscious "as vulnerabilities to exploit" by bypassing our capacity for conscious thought.[86] What they get wrong is that they build on an outdated Freudian view that our psyche has "two minds"—a conscious and an unconscious one. We have since learned that unconscious processes use the same brain regions in the same ways as conscious processes. Our unconscious mind is primed *all the time* through regular "strength" stimuli (rather than hidden and subliminal ones). Think of the popcorn and soda advertisements before a movie begins. They are hardly hidden, but they play to our baked-in desires. Advertisers and tech giants have just gotten much better at identifying and targeting them. Indeed, social psychologists have argued for decades that people are unaware of the powerful influences that are brought to bear on their choices and behavior.[87]

Tanya Chartrand, a Duke professor, provides a different way to understand the problem by distinguishing between different kinds of unawareness—of primes that triggers our mental process, of the mental process itself, or of the consequences and effects of those triggers. Yale professor of psychology and cognitive science, and professor of management John Bargh, who has long studied unconscious influences on the consumer mind, argues that it's the last of these we focus on—"not whether the event itself is perceived consciously or not (it almost always is), [but] whether the person is aware of how that event *affects* their choices and behavior."[88] Poets, politicians, governments, and advertisers all know how to exploit the unconscious influences on our lives, he explains. And advertisers know well how to manipulate hidden mechanisms that drive our behavior. But people are often unaware of how those influences impact their actions. Which is why it's critical that we understand what others can and can't do to change our minds.

In chapter 8, we'll consider the starkest examples of manipulation perpetrated by assaulting our brains, which clearly violate our right to self-determination and freedom of thought. The more difficult cases to resolve, however, are the subtler influences that shape our everyday decision-making and that are quickly becoming normalized.

It's much easier to prime us to act in ways that are consistent with our existing goals. Advertising a weight-loss program to a person who is trying to gain weight will inevitably fail. Priming us with cues that are related to our goals, however, will focus our "selective attention" on "goal-relevant features of the environment," which can shape our choices that follow.[89] Professors of marketing and psychology Gráinne Fitzsimons, Tanya Chartrand, and Gavan Fitzsimons found compelling evidence of this effect when they subliminally primed study participants with Apple and IBM brand logos. The Apple logo prime led people to act more creatively on subsequent study tasks compared to subliminal IBM logo priming—but only when creativity was a part of the participants' self-descriptions.[90] Apple evoked in these participants an association of creativity, leading those with a prior stated goal of being creative to act more creatively on subsequent tasks. Because IBM didn't evoke the same association, even those with creativity as a stated goal didn't act more creatively when primed with IBM instead.

Subliminal priming can also nudge us to act in ways that are irrelevant to goal-directed behavior. Bargh describes a study in which participants had been unknowingly primed with words related to betting (bet, gamble, wager) or standing pat (pass, fold stay) on given hands in games of online blackjack. The participants' betting was statistically consistent with the primes they had been given. But they believed that they had freely and consciously chosen their bets—and the belief was even stronger when they were primed than when they were unprimed.[91]

Even asking us questions about our hidden vices can change our subsequent behavior.[92] We often have conflicting attitudes about behaviors like smoking, drinking, and using drugs. We get a short-term reward (like a dopamine hit in our brain) when we indulge, but we also understand the negative long-term consequences that go with them. When we hold conflicting explicit negative and implicit positive attitudes about a behavior, priming may give us "license to sin."[93] Frankfurt's human addict wants to break his addiction but asking him how often he plans to take the drug in the next week can nudge him toward doing so more often, despite his explicit preference otherwise.[94] When researchers asked students about their attitudes toward skipping class, they reported strongly negative attitudes toward doing so, but then skipped class more frequently in the weeks following.[95] When study participants were asked how often they would go out drinking or watch television instead of studying, they also did so more frequently in the week following.[96] But when framed negatively—telling participants that drinking and wasting time watching television are vices to be avoided—the vice behavior remained the same.[97] How an influencer frames a question can liberate us to sin or increase our ability to avoid doing so.

All of which makes it exceptionally unrealistic at best, or outdated at worst, to define unlawful manipulation as intentionally using hidden influences to affect our decision-making. When neuromarketers use advances in neurotechnology to discover what makes us tick, and then use that information to make their products more enticing, they don't render us unable to act consistently with our goals. As of yet, no one has discovered the so-called buy button in our brains. When the Disinformation Dozen exploit evolutionary shortcuts in our brains to make us more susceptible to fake news, they don't prevent us from getting vaccinated, even if their bad arguments do appeal to our heuristics.

But when Nir Eyal teaches companies how to addict us to their products, despite the long-term negative consequences of doing so, we should make sure we retain the ability to act otherwise and investigate whether they intend or cause actual harm as a result. If a product becomes actually or nearly impossible to resist, our freedom of action will be hindered and our self-determination and freedom of thought will be put at risk.

Dr. Shaheed concedes that freedom of thought cannot and should not be used to prevent "ordinary social influences, such as persuasion." We may encourage others, advise them, even cajole them, he argues. But at some point, an influence crosses the line from permissible persuasion to impermissible manipulation.[98] He offers a nonexclusive set of factors to consider, including whether the person has (1) consented to the practice with fully and freely informed consent; (2) if a reasonable person would be aware of the intended influence; (3) if there is a power imbalance between the influencer and target; and (4) if there was actual harm to the person subject to manipulation.[99]

These are helpful but still don't make clear the *nature* of the influence we are defending ourselves against. We can't and shouldn't attempt to regulate every marketer, politician, artist, or entity who tries to appeal to our unconscious biases, desires, and neural shortcuts, lest we interfere with everyday interactions that are part of what it means to be human, whether those attempts are hidden or visible, or targeted at our unconscious or conscious neural processes. But when a person or entity tries to override our will by making it exceedingly difficult to act consistently with our desires, and they act with the intention to cause actual harm, they violate our freedom of action, and our right to cognitive liberty should be invoked as a reason to regulate their conduct.

However begrudgingly, we must admit that neuromarketing per se does not violate cognitive liberty, so long as the research is conducted ethically and the findings are not used to intentionally cause us harm. We can't say the same about intentional efforts to exploit our brains by addicting us to technology, social media platforms, or other products. While our brains may fall for bad arguments when cleverly framed, we can and should encourage societal interventions that nudge us to slow down and think critically. When Twitter asks "Would you like to read the article first?" before retweeting it, it's asking us to slow down and think critically before we act. More companies ought to implement mechanisms that encourage users to do the same. And we should aspire to do so ourselves even when we aren't nudged to do so. But freedom of thought shouldn't be used as an excuse for filtering that information for us.

As for Dr. Shaheed's recommendation that we consider whether a person has freely and voluntarily consented to an intervention? While consent will rarely be enough to shield us from the coming encroachments to cognitive liberty, with at least the newest technique we turn to next, it should be a critical factor in considering the legitimacy of the technique.

Bypassing the Conscious Brain

A few months ago, I had a fitful night of sleep plagued by intrusively vivid dreams in which I was consoling a friend. I woke up in a state of agitation. Suddenly my phone buzzed with a notification. It was that same friend—whom I had not spoken with for many months—asking if I was okay. "It's so strange to hear from you this morning," I replied. My phone dinged again. "I had an incredibly vivid and disturbing dream about you," my friend continued. "I dreamt you had died. I was attending the memorial service. I was inconsolable." I was deeply shaken. Not just by the idea that my friend had dreamed I was dead! I was much more disturbed by the seeming synchronization between our dreams. What could that possibly mean? Was it just a random coincidence? (I rarely dream about this friend, and suspect my friend rarely dreams about me.) I then had an even more disturbing thought: What if it wasn't a coincidence at all? What if our dreams had somehow been tampered with? That may be less far-fetched than you think.

For years, Coors has been locked out of showing commercials during the NFL Super Bowl game, because the league has a contract with Anheuser-Busch. So, Coors found a new way to infiltrate the minds of NFL onlookers. With help from Dr. Deirdre Barrett, a psychologist who is an expert on dreams, it created a film with specific audio and visual stimuli. In exchange for half off a twelve-pack of Coors beer (or a free twelve-pack if shared with a friend), Coors invited participants to watch a ninety-second video featuring images of mountains and Coors beer right before going to bed, and then listen to a soundscape while they were sleeping. When they woke up, they were asked what they had dreamed about.[100] Dr. Barrett claimed the "participants reported similar dream experiences, including refreshing streams, mountains, waterfalls, and even Coors itself."[101]

How could this possibly work? Upon awakening from a dream, we have a period of about twenty minutes before blood flow is fully reestablished to the dorsolateral prefrontal cortex, during which we are more suggestible. Dr. Barrett and other researchers capitalize on this to incubate dreams. Suppose you decide, for example, that you want to work out some emotional turmoil you've been experiencing while sleeping. You could record a dream prompt message for yourself using your own voice. That prompt will then be played back to you as you fall asleep. After falling asleep, a sleep sensor would wake you up and play your recording to guide your thoughts during the time your brain is in a suggestible state. You would then fall back asleep only to be awakened again after a predetermined time and prompted with the same dream message. This cycle of sleep and re-prompting would continue throughout the night to guide your dreaming.[102]

Coors is not the only company that is using dreams to sell things. Xbox's "Made from Dreams" was created to promote the Xbox Series X. The videos were crafted from dreams selected streamers reported after playing the new console for the first time.[103]

This seeming sci-fi scenario is made easier by sensors that can detect brain wave activity, pinpointing the stages of sleep we are in at any given moment.[104] Even a bedside smart speaker can be used to detect the breathing patterns indicative of different stages of sleep, triggering it to play soundscapes that can incubate dreams.[105] Researchers have even communicated directly with lucid dreamers—the stage of sleep where one is aware while dreaming—and had them answer questions and solve math problems.[106]

There is great potential to these advances—to treat nightmares, enhance learning and memory consolidation, even overcome PTSD and addiction.[107] But dream researchers also caution about the perils of doing so. Collectively, they have started to author ethical norms and guidelines because of "the threat of the capture, sale and colonization of the dreaming self in the form of data; the outsourcing of introspection, trusting sensors more than senses; and the infiltration of our most private spaces by those who may wish to harm or manipulate us."[108]

In June 2021, forty sleep and dream researchers published an open letter in which they stated that targeted dream incubation "is not some fun gimmick, but a slippery slope with real consequences."[109]

One of its lead authors, cognitive scientist Adam Haar, had invented a device that tracks sleep patterns and guides wearers to dream about specific subjects by playing audio cues. He was alarmed when companies ranging from big tech to major airlines started to ask him for help incubating consumers' dreams. He worries that people are particularly vulnerable to suggestive content when sleeping and fears that the absence of regulations for in-dream advertising could lead to a future in which we "become instruments of passive, unconscious overnight advertising, with or without our permission."[110]

But other researchers, like Tore Nielsen, a dream researcher at the University of Montreal, are far less troubled by those dystopian scenarios, because the interventions are unlikely to work unless the dreamer is aware and willing to participate in the dream incubation. "I am not overly concerned," he said, "just as I am not concerned that people can be hypnotized against their will."[111]

I didn't agree to dream about my friend that night. I suspect it was just a random coincidence, or that something had unconsciously primed us both to think about each other earlier that day. But the experience did remind me of the coming possibilities and reaffirm in my mind that at the very least, dream incubation is an intervention that ought to be consented to anytime it is used. It's one thing to target our unconscious processes when our conscious processes are active and can be used to defend our freedom of thought. But it's much more troubling to have our unconscious processes infiltrated when our conscious processes are silenced during sleep and cannot be used to filter information. Nonconsensual dream incubation—whether used to benefit or harm us—falls much closer to the wrong end of the spectrum between innocent influence and intentionally assaulting the brain, the practice we turn to next as a stark violation of cognitive liberty.

8

Bewilderbeasts

It was the fifth meeting of the Presidential Commission for the Study of Bioethical Issues, and the main item on our agenda was the adequacy of existing federal standards for research on human subjects. The ballroom at the Warwick New York Hotel had banquet-style chairs lined up in rows for the public. Every seat was filled, and there were standees in the aisles and along the sides of the room.

The thirteen commissioners sat around a U-shaped table. Our chair, then president of the University of Pennsylvania, Amy Guttman, and vice-chair, then president James Wagner, of Emory University, sat next to each other where it curved, at the head of the room. I was seated at the end of one of its arms, close to the audience. It consisted of people mostly dressed in conservative business attire and seemed to radiate an almost palpable energy.

We had posted a notice in the Federal Register inviting public commentary and had received more than three hundred submissions in advance. Only a few would have time to speak in the allotted hour for public comment, and each was given a minute and a half to address the commission. One at a time, they approached the microphone.

The first to speak was a middle-aged Black woman, dressed in a conservative long-sleeved gray sweater, with a colorful scarf tied around her neck. Squinting through her glasses at the prepared statement shaking in her hand, she read it quickly and passionately. "I'm a targeted individual," she began. "Gangstalking and harassment, nonconsensual biotechnology application is being used on me." I glanced worriedly at a fellow commissioner.

"I now experience involuntary limb movements," she continued. "I receive stingings. I get pains to my head . . . I get ringing in my ears . . . I get pulsing sensations in my body. I get an electrical current, an electrical sensation that goes up and down through my body and can be isolated to different parts of my body . . . I feel as though I am being roboticized."[1]

She invited others in the room who shared her experience to stand up, and suddenly half the audience was on their feet. A second person approached the microphone. This one had an unkempt mane of gray hair, wire-rimmed glasses, and he was squeezed into a suit that was a little too small on him, with a powder-blue button-down shirt underneath. Midway through his remarks, he broke down in tears.

"I've been a victim of ongoing nonconsensual human subject experimentation for my entire adult life," he said, "and possibly may have been a victim since my childhood. I have been targeted with ongoing microwave weapons, as well as drugging with neurotoxic contaminants covertly placed on articles of clothing, as well as on other personal possessions."[2]

One after another they spoke, as our vice-chair compassionately moderated, validating their right to be heard and the importance of what they had to say. I became fearful for my safety toward the end, as one man aggressively approached the microphone and demanded to know whether our chair had received his letter and reached out to President Obama as he'd asked.

Growing numbers of people around the world believe that the government, their employers, and even their neighbors and friends are subjecting them to constant surveillance, harassment, and mind control. These self-described targeted individuals, or TIs, have their own vocabulary—google "targeted individual" or "gangstalking," and you'll enter their world.

While investigating the proliferation of their claims in online forums,[3] Amelia Tait, a reporter for *MIT Technology Review,* turned to psychologist Lorraine Sheridan and stalking expert David James for insights. Sheridan and James had recently published an article detailing the results of an anonymous survey of 128 people who claimed to be TIs subjected to gangstalking, a term that has emerged to describe a feeling of being stalked and harassed by a group of people.

Tait and Sheridan sorted their claims into three buckets: (1) "cases where the resources or elaborate organization required to carry them out made the alleged activities highly improbable"; (2) "cases in which the activities described were impossible (e.g., minds of friends and family being externally controlled; use of 'voice to skull' messages . . . insertion of alien thoughts; organized electronic mind interferences. . . . Invasion of an individual's dreams at night)"; and (3) "cases where the beliefs were not only impossible, but bizarre (e.g., . . . remote enlargement of bodily organs)."[4] All of them "were highly likely to be delusional in nature" and "indicative of paranoid delusional systems."[5]

Joel Gold, a professor of psychiatry at New York University School of Medicine and coauthor of *Suspicious Minds: How Culture Shapes Madness,* explains how as technology evolves, people with delusions incorporate those advances into their delusions. Given the ubiquity of digital tracing and closed-circuit video cameras, he says, it's natural to believe we are under constant government surveillance. Modern technologies "are the seeds of reality that people use to build their delusion upon."[6] Those seeds of reality—real technologies and government programs with aims to surveil and control people—get incorporated into common narratives thanks to online forums where people gather and share their research, beliefs, and perceptions with one another.

We beefed up security for subsequent meetings and invited public comments in writing only. While individuals who believe they are being persecuted are more likely to spiral into depression and isolation than violence,[7] it is not uncommon for them to perceive hostility in social situations, which on very rare occasions can have deadly results.[8]

Aaron Alexis, a thirty-four-year-old US Navy contractor who said he had been hearing voices and was being harassed through microwave mind control, shot and killed twelve people and injured three others in the Washington Navy Yard in 2013.[9] Gavin Long, a former marine, carefully planned an ambush that he carried out on his twenty-ninth birthday, killing three law enforcement officers and injuring three others in Baton Rouge, Louisiana, in 2016, after posting messages and videos online claiming he was under constant surveillance.[10] Myron May, previously a successful lawyer, asked in an online forum whether others had been "encouraged by your handler to kill." After leaving voice-mail messages in which he said he was being attacked by energy weapons, and that he would expose his persecutors "once and for all," he opened fire in the Florida State University Library, killing three people.[11] Semmie Williams posted thousands of social media messages and videos claiming he was being followed, mocked, and sexually assaulted, and that police were targeting him with electromagnetic field weapons.[12] In December 2021, he was arrested and charged with the murder of Ryan Williams, a teenager from Palm Beach Gardens. His public defender says that Williams suffers from "long-standing and persistent mental illness."[13]

In recent years, several self-identified TIs from all walks of life—doctors, lawyers, military personnel, artists, and others—have presented their claims before legislatures, government commissions like ours, and in courtrooms, hoping to raise public awareness about what they believe is an international human rights issue.[14] As an advocate for human rights protections to secure our mental experiences, I often hear from them. While I am never quite sure how to respond or what I can do to help, their pleas keep coming.

It is against this background, and only with great trepidation, that I will examine the "seeds of reality" that underlie a handful of their claims. As paranoid as it might sound, and as troubled as most self-described TIs undoubtedly are, at various times and places throughout history, governments have indeed attempted to develop mind-control capabilities—and some of those efforts have continued to this day. Because of their capacity to rob of us our personhood, agency, and very capacity to make choices, they are among the starkest threats to our self-determination and freedom of thought.

For most of the topics I've covered thus far, there have been shades of gray; our cognitive liberties are not always absolute. But when it comes to the weaponization of mind control, virtually every example is clearly over the line.

MK-Ultra and Mind Control

I'm usually checked out when watching (yet another) animated film with my kids, but *How to Train Your Dragon 2* had me riveted. Toothless, the lead dragon, had succumbed to mind control by the alpha dragon, Bewilderbeast. Toothless eventually breaks free, but not before he is forced to kill a human. My daughter Aristella was heartbroken, and I was shaken as I connected what I was seeing with what I was learning about governments' modern and historical attempts to become Bewilderbeasts themselves.

In 1953, as the Korean War was coming to an end, a number of startling stories about American prisoners of war (POWs) emerged. Some had publicly confessed to crimes that the United States categorically denied—like carrying out biological warfare. Others became "turncoats" and refused to return to the United States.[15] The CIA was convinced that they had somehow been "brain-washed" by the Communists, via a combination of drugs, behavior modification, and torture.[16]

Addressing a gathering of Princeton alumni on April 10, Allen Dulles, the newly appointed CIA director, spoke out vehemently against suspected "Soviet brain perversion techniques," calling them "abhorrent" and "nefarious." He worried that war had become as much an ideological "battle for men's minds," and that the West was handicapped when it came to "brain warfare," because mind-control was so inimical to our core values.[17] He set out to change that.

Three days later, Dulles launched the top secret CIA MK-Ultra program, earmarking $25 million so the United States could develop its own "mind-control" capabilities.[18] Marie D. Jones, coauthor of the book *Mind Wars: A History of Mind Control, Surveillance, and Social Engineering by the Government, Media, and Secret Societies,* described MK-Ultra's ultimate goal as the discovery of a way to erase "the subconscious of a victim and replac[e] [it] with a new way of thinking."[19]

The CIA believed the Soviets were using lysergic acid diethylamide, or LSD, on American POWs.[20] So, Sidney Gottlieb, a biochemist who had been leading the CIA's own search for a "truth serum," arranged for the CIA to buy up the then-existing world supply of LSD and planned a series of experiment to unlock its mysteries [21]—some using volunteers, some in which the drinks of CIA employees were spiked without their knowledge.[22]

MK-Ultra went on to include at least 144 different subprojects, carried out across 89 known institutions, including some major universities.[23] In its quest to find the most effective means of hacking the brain, the program inflicted electroshock therapy, hypnosis, polygraphs, radiation, drugs, toxins, and chemicals on willing and unwilling subjects alike. "Patients at psychiatric hospitals, prisoners in federal institutions, and even people in the public were given drugs without their awareness or consent and experimented upon," according to the investigative journalist and historian Tom O'Neill."[24] CIA officers in Europe and Asia used these techniques on captured spies.[25]

Some of the most devastating experiments took place at the Allan Memorial Institute in Montreal, Canada, where the Scottish-American psychiatrist Dr. Ewan Cameron used an aggressive combination of customized drug cocktails and extreme "depatterning" techniques on his unsuspecting patients. This "depatterning" treatment was meant to return them to an infantile psychological state so he could rebuild their minds. He began by inducing an insulin coma in the patient, then exposed them to audio recordings of messages like "Your mother hates you," which were repeated hundreds of thousands of times. Alternately, he used Page-Russell shock therapy, administering electrical currents at forty to seventy-five times the recommended strength to wipe their memories clean. We don't know how many of these subjects died, but we do know that those who survived were never the same. Many suffered profound personality changes, lost their memories, or became uncontrollably violent.[26]

The CIA knew even then that these studies were deeply unethical and intentionally hid them from public view.[27] With telling irony, they brought in German and Japanese war criminals and vivisectionists and invited them to share what they had learned from their own experiments on prisoners.[28]

When John Vance, a member of the CIA inspector general's staff, learned about the use of "unwitting nonvoluntary human subjects" in 1963, he insisted MK-Ultra be ended or brought in alignment with current ethical norms. By this time, anxieties about brain control had become widespread. Michael Wood, a lecturer at the University of Winchester's Department of Psychology, credits the 1962 film *The Manchurian Candidate*, a story about a soldier manipulated by mind control into killing a politician, for bringing MK-Ultra out of the shadows and into pop culture.[29]

The CIA steadfastly denied the program's existence until Congress investigated in 1977.[30] By then, much of the evidence had been deliberately destroyed.[31] Most of what we know about MK-Ultra today comes from several boxes of records that had been overlooked, as well as the scant CIA testimony before Congress.[32]

"MK-Ultra sounds so cartoonish, almost like the dastardly scheme of a Bond villain," says Wood, "but its origins are based on verifiable facts and that gives it an uncomfortable edge."[33] An American six-part docudrama miniseries, *Wormwood*, released on Netflix in 2017, tells the true story of the bacteriologist and biowarfare expert Frank Olson, who fell to his death from a hotel window after unknowingly being dosed with LSD. In May 2022, Cinedigm acquired North American rights to the thriller *MK Ultra*, about a psychiatrist who finds himself in an ethical quandary when he is recruited to oversee part of the program.[34]

MK-Ultra is also, of course, one of the critical "seeds of reality" that animates modern conspiracy theories. When the rapper Cardi B was caught on camera staring into space during a red-carpet interview at the 2018 Grammys, conspiracy theorists leaped on it as modern proof of ongoing MK-Ultra mind control.[35] Online forums are rife with MK-Ultra claims. Mass shooters are frequently identified as "MK-Ultra puppets" acting under mind control.[36] Marie Jones worries that the assimilation of MK-Ultra into popular culture may numb us to the stark reality of what transpired, trivializing the work of "people [who are] seriously studying the history of MK-Ultra," and making it that much likelier that the same abuses will be committed again.[37]

Modern-Day War on the Brain

The past is already present in China. In late 2021, the US government blacklisted twelve Chinese institutes and firms it believed to be working on dangerous "biotechnology processes to support Chinese military end uses," including "purported brain-control weaponry."[38] This followed reports that the Chinese People's Liberation Army (PLA) is investing heavily in warfare techniques for "cognitive domain operations,"[39] including $85 million allocated by the PRC's Ministry of Science and Technology in 2020 to fund AI research, and "brain-inspired software and hardware, human-machine teaming, swarming, and decision making."[40] A recent US Pentagon report describes the growing focus by the Chinese on intelligent warfare, as well as intelligent swarms (such as drones that move collectively in response to a single command), AI-based space confrontations, and cognitive-control operations.[41]

Nathan Beauchamp-Mustafaga, a China specialist at the RAND Corporation, calls this nothing less than an "evolution in warfare, moving from the natural and material domains—land, maritime, air, and electromagnetic—into the realm of the human mind."[42] The PLA, he says, hopes to "shape or even control the enemy's cognitive thinking and decision-making" abilities.[43] From disinformation campaigns to modern weapons targeting the brain, Dulles's "brain warfare" is rapidly becoming a reality.

Translated Chinese military reports obtained by the *Washington Times* further support these reports. "War has started to shift from the pursuit of destroying bodies to paralyzing and controlling the opponent," said one, titled "The Future of the Concept of Military Supremacy."[44] This is in accord with publications by Chinese National Natural Science Foundation–funded scientists, who have urged China to invest in military brain science (MBS). The brain, they argue, is the "'headquarters' of the human body," and weapons "[p]recisely attacking the 'headquarters'" will soon become "one of the most effective strategies for determining victory or defeat in the battlefield."[45] More research is needed, they say, into "the damaging effects to sensitive target areas of the brain tissues by acoustic weapons, laser weapons, high-explosive weapons, and electromagnetic weapons." "Brainwave interference" and "infrasound weapons," they say, are being developed that "interfere with brain tissue and cause insanity through resonance."[46] "Interfering with the brain," they write, "can affect mentality, influence thinking, and affect decision making to create a whole new 'brain war' combat style and redefine the battlefield."[47]

Like other countries, the United States is investing heavily in military applications of brain–computer interface (BCI) technology in the hope of creating super soldiers that can control swarms of drones with their minds, communicate and upload data brain-to-brain, and identify targets subconsciously.[48] But BCI is extremely vulnerable to hacking, and Chinese scientists point to the military potential for doing just that.[49]

Gina Raimondo, the US commerce secretary, recently expressed concern that China will use its developing capabilities in cognitive control on its own citizens—including Uighur Muslims, more than a million of whom are being held in camps in Northern China.[50] "Unfortunately, the People's Republic of China is choosing to use these technologies to pursue control over its people and its repression of members of ethnic and religious minority groups," she cautioned.

China has called the US sanctions "unwarranted suppression" that violate free trade. The spokesperson for the Chinese embassy in the United States, Liu Pengyu, described the projects in question as designed for the well-being of China's people.[51] But persistent claims of unexplained brain injuries from diplomats around the world could make anyone wonder whether it's just China's citizens who are in the crosshairs of these frightening new technologies.

Losing Our Minds

One of the most insidious things about losing your mind is that other people can't see when it happens. When I am debilitated by a migraine, no one but me is aware of the pain I am experiencing. But sometimes, even we don't realize when our minds are no longer our own.

When Barbara Lipska, a neuroscientist and director of the Human Brain Collection Core at the National Institute of Mental Health in Bethesda, Maryland, started to lose her mind, she didn't connect the dots between what she was experiencing and what she had spent most of her lifetime studying. In her memoir *The Neuroscientist Who Lost Her Mind,* she remembers one ordinary morning in 2015. She got up and dyed her hair, then she went out for a run.

When Lipska returned much later than typical for her, her husband was shocked by her appearance. "The hair dye that I put in my hair that morning dripped down my neck. I looked like a monster,"

Lipska said. Over the next two months, Lipska experienced symptoms that looked a lot like dementia and schizophrenia. In an interview with NPR, she described how she changed "from a loving mother, grandmother, and wife, into a kind of heartless monster. I was yelling at my loving husband. I was yelling at my beloved grandsons and my children. I was behaving like a two-year-old with a tantrum—all the time."[52] What she didn't realize is that her mind was under assault, in her case, by the cancerous tumors that were growing in her brain. Once diagnosed, and treated, she was one of the lucky people who recovered fairly quickly. Despite some lasting effects—such as a loss of vision, balance, and spatial orientation—she regained full control of her mind.

Mind control provides fertile ground for narrative nonfiction and science fiction alike. Of the many depictions of mind control in science fiction, a few stand out for me: Robert A. Heinlein's 1951 classic, *The Puppet Masters*, in which US intelligence officers fall prey to sluglike telepathic aliens; Anthony Burgess's *A Clockwork Orange*, with its Ludovico Technique—a form of mind control that makes its antihero, Alex, feel pain whenever he has a violent or antisocial impulse; and probably most important in my life and thinking, George Orwell's *Nineteen Eighty-Four*, which describes both direct and indirect mind control through the artificial language of Newspeak.

Or who could forget the powerful scene in J. K. Rowling's *Harry Potter and the Goblet of Fire*, when Harry is subject to the Cruciatus Curse, also called the Torture Curse, and Voldemort takes hold of his mind:

> Voldemort raised his wand, and before Harry could do anything to defend himself, before he could even move, he had been hit again by the Cruciatus curse. The pain was so intense, so all-consuming, that he no longer knew where he was . . . white-hot knives were piercing every inch of his skin, his head was surely going to burst with pain; . . . . And then it stopped . . . "A little

> break," said Voldemort, the slit-like nostrils dilating with excitement, ". . . that hurt, didn't it, Harry?"

These real-life and fictional stories force us to grapple with the terrifying possibility of losing our minds. When you add to this common anxiety the reality of governments' past, present, and future weaponization of cognitive control, the people who spoke before our committee don't seem quite so delusional.

Recent claims by US service personnel make it plausible that the thing we fear is already here.

Scores of US diplomats stationed in Cuba and China started reporting in 2016 that they were experiencing strange mental ailments including recurrent headaches, blurred vision, vertigo, and hearing strange sounds. Scientists and the intelligence community have been stumped on its cause. Deliberate physical attacks, involving unknown brain-control technology? Psychogenic symptoms, or subliminal attacks on the human mind? The acute onset of the symptoms, the consistency across so many of the reports, and the high-level government personnel involved have led some to conclude these reports can only be explained as a coordinated attack by an unknown adversary.[53]

Olivia Troye, former homeland security and counterterrorism advisor to Vice President Mike Pence, was one of the victims. She had previously served in the Pentagon, in Iraq, in the Defense Intelligence Agency, and at the National Counterterrorism Center. In the summer of 2019, while she was working at the White House in Washington, DC, she experienced a sensation that felt like a physical blow to her head while climbing a stairway in the West Wing. Describing it on *60 Minutes*, she said felt "this piercing feeling on the side of my head, it was like, I remember it was on the right side of my head and I got like, vertigo. I was unsteady, I was, I felt nausea, I was somewhat disoriented, and I was just, I remember thinking, 'Okay, you gotta— don't fall down the stairs. You've gotta find your ground again and steady yourself.'"[54] About a year later, she experienced the same thing while walking to her car.

A senior member of the National Security Council claimed he suffered a similar attack while on the exact same stairway. He couldn't speak or think clearly and was taken to the ER for treatment. Miles Taylor, deputy chief of staff and later chief of staff of the Trump administration's Department of Homeland Security, claimed he was hit in late April 2018. He described waking up in the middle of the night in his apartment to an eerie sound, "a sort of chirping, something between a cricket and a digital sound."[55] About five weeks later, it happened again, leaving him feeling as if he had suffered a concussion. Dozens of Americans overseas reported hearing a similar sound before experiencing acute symptoms of their own.[56]

By June 2021, more than 130 US officials in Cuba, China, Russia, Colombia, Austria, the United Kingdom, and more had reported similar events.[57] By 2022, the intelligence community had surmised that some unseen weapon was being deployed against Americans' brains.[58] The reports were taken so seriously that they prompted rare bipartisan action by the US Congress to unanimously pass legislation, signed into law by President Biden, to secure compensation and medical care to victims with confirmed brain injuries.[59]

The US Department of State turned to the National Academies of Sciences, Engineering, and Medicine (NASEM) to consider the possible sources of the injuries, which in turn created the Standing Committee to Advise the Department of State on Unexplained Health Effects on US Government Employees and Their Families at Overseas Embassies. While the NASEM Committee's investigation was hampered by lack of access to the service personnel and classified information, they nevertheless found a unique set of clinical signs and symptoms that didn't fit the pattern of any existing neurological condition.

What was distinct about these reports was the nature and onset of the initial symptoms: "the sudden onset of a perceived loud sound, a sensation of intense pressure or vibration in the head, and pain in the ear or more diffusely in the head."[60] These features made it difficult to simply blame psychological or social factors, even though some of the differences between the claims might be explained by them. Though clinicians at the University of Pennsylvania, to whom the US Department of State had directly referred personnel and family members for evaluation following potential exposure while serving in Havana, Cuba (hence the name Havana syndrome), found evidence of brain damage,[61] the Standing Committee didn't believe the evidence was clear enough to medically classify the phenomenon. Ultimately, it concluded that a directed radio frequency (RF) energy attack was the most plausible explanation. The acute onset of symptoms made chemical exposure unlikely, and the symptoms themselves were "consistent with RF effects."[62]

Other scientists were deeply skeptical of the committee's findings. Cheryl Rofer, a former chemist at the Los Alamos National Laboratory, found it implausible that a microwave weapon was involved. "Aside from the reported syndromes, there's no evidence that a microwave weapon exists—and all the available science suggests that any such weapon would be wildly impractical," she wrote in *Foreign Policy*.[63] After the University of Pennsylvania published its findings of brain damage in 2019, fifteen eminent neuroscientists and physicists published a letter calling the work "deeply flawed" with "poorly founded conclusions" that were likely the result of political pressure.[64] Other scientists attributed the reports to mass psychogenic illness, arguing that directed energy weapons are in their earliest stages of development, and thus incapable of achieving the kind of targeted injuries that victims claim.[65]

Medical sociologist Robert Bartholomew has written multiple books about mass psychogenic illnesses throughout history, citing more than 3,500 different cases over time. You can "make yourself feel sick if you think you are becoming sick. Mass psychogenic illness involves the nervous system, and can mimic a variety of illnesses," he explains. Take "Bin Laden Itch."[66] Thousands of US schoolchildren broke out with itchy red patches between 2001 to

2002, after a Florida man was diagnosed with anthrax poisoning, and the nation was gripped by fears of terrorism following the unprecedented September 11, 2001, attacks. The rashes disappeared without further progression, and there was never a medical explanation for the cause. This is a classic example of mass psychogenic illness, Bartholomew explains, which is not something "all in the head" but produces real physiological symptoms. He thinks the simplest and most logical explanation for Havana syndrome is the same, because it has the classic hallmark of the phenomenon: starting with high-status opinion-formers, spreading among a closely knit group of individuals, occurring in a high-stress environment, colored by the power of suggestion.

A recently declassified summary of the findings of an expert panel convened by the Biden administration, however, agreed with the National Academies' findings, despite the lack of physical evidence in support of an RF weapon.[67] The panel was established to identify potential mechanisms behind what are now being called Anomalous Health Incidents affecting US government personnel. It examined thousands of classified documents and interviewed some of the individuals who had reported symptoms, focusing on five potential causal mechanisms: acoustic signals, chemical and biological agents, ionizing radiation, natural and environmental factors, and radio frequency and other electromagnetic energy.[68]

Curiously, many sections of the declassified report were redacted, including the entire section on how pulsed electromagnetic energy would work and the kinds of antennas that would be needed to propagate such a signal. The entire next paragraph was also redacted, which specified the way that such a weapon would be deployed. The panel was not tasked with, nor did they opine on the question of "whether a foreign actor may be involved."[69]

It came as a surprise to many in the intelligence community when, in January 2022, the CIA issued an interim report implicitly agreeing with Bartholomew by attributing all but about two dozen of the claims to stress. A CIA official who spoke on condition of anonymity admitted "We have so far not found evidence of state-actor involvement in any incident," although the agency has "not ruled out the actions of a foreign actor."[70] This was news to a number of senators and congressmen on intelligence committees, who had seen information that pointed toward the use of directed-energy weapons.[71] Jim Giordano, a senior fellow in biotechnology, biosecurity, and ethics at the US Naval War College, who was brought in as a government advisor after diplomats fell sick, told *The Guardian* that he believes that Russia has continued Soviet-era research into directed-energy weapons, and that China already has those capabilities.[72]

Both the mystery and the investigations continue. In May 2022, a senior Biden administration official who spoke at a White House briefing said "we are working across the interagency and State Department and with allies around the world as well to get to the bottom of the anomalous health incidents. And at this moment, we do not have a conclusion as to the attribution."[73]

Has China or Russia perfected and deployed electromagnetic weapons? Or is this a contagion of psychogenic illnesses? The crucial issue is that these weapons are either being developed or have already been deployed.

Self-Determination and the Capacity for Choice

In November 2020, François du Cluzel, a project manager at NATO ACT Innovation Hub, a community where experts and innovators collaborate to tackle NATO challenges, issued a report titled *Cognitive Warfare,* which identified the human domain as the next battlefield frontier. The nations of the world are in a race to weaponize neuroscience, he wrote.[74] Which is why he believes that NATO should already be working to establish a common understanding of when and how cognitive weapons can be deployed in a manner consistent with law and accepted international norms, and how the Law of Armed Conflict would or should apply to the use of cognitive technologies by military forces in the future.[75] To do this, he argued, we will have to update our understanding of customary international law, and the treaties that govern conduct during war, including the Geneva Convention and law of the Hague.

The weaponization of neuroscience potentially robs individuals of the very principles upon which international human rights law is founded—the principles of dignity and agency. Thus far, we have grappled with the right to self-determination over our brains and mental experiences, concluding that our right to self-determination includes the right to be free from societal interference with our brains unless our choices directly affect the interests of others. That right must also include freedom from interference with our *capacity* to make choices—a freedom that may, under limited circumstances in armed conflict, be subject to exceptions, but not in times of peace, and never to prosecute undeclared wars against people's minds.

Targeting brains to destroy, coerce, or control people's mental experiences compromises human agency and dignity and defeats our right to self-determination. Doing so also violates our freedom of thought by robbing us of the ability to think freely—the most drastic form of manipulation we have encountered. Weapons targeting the human brain may render us unable to make our will effective or strip us of our capacity for freedom of will altogether. While we cannot and should not attempt to regulate every appeal to our unconscious brain processes, attempts to seize, disable, and override human agency fall well outside those bounds.

To fill the void that Du Cluzel identified, we should look to international human rights guidance on freedom of thought and psychological torture, including the guidance recently issued by Nils Melzer, the former UN special rapporteur on Torture and Other Cruel, Inhuman or Degrading Treatment. In his March 2020 report to the UN General Assembly, Melzer argued that states are using psychological torture to circumvent the more widely accepted ban on the physical infliction of pain that was adopted in 1984. Although his report was primarily focused on other means of bombarding the mind, he nevertheless noted that advances in neuroscience could be exploited to inflict "invisible" pain via psychological torture.[76]

Psychological torture has long been treated as analytically distinct from other forms of torture.[77] It shouldn't be. It can inflict just as much suffering as other forms of torture and dehumanizes a person by robbing them of their personhood. Article 1 of the UN Convention Against Torture and Other Cruel, Inhuman or Degrading Treatment or Punishment defines torture to include "severe pain or suffering, whether physical or mental . . ."[78] But the tools of psychological mind control don't always inflict severe pain; allowing actors to circumvent the Covenant's restrictions. But expanding our interpretation of "suffering" in international law to prohibit debilitating a person's mind would eliminate the gray area that governments have exploited.

From solitary confinement to coercive interrogation, psychological torture has been deployed as a tool of governments across the world and across time. Given the rapidly growing focus on cognitive warfare, Melzer recommends that states incorporate definitions of psychological torture as a subcategory of torture, "to include all methods, techniques and circumstances which are intended or designed to purposely inflict severe mental pain or suffering without using the conduit or effect of severe physical pain or suffering."[79]

His report cites historical "mind control" experiments as well as "new and emerging technologies" that "give rise to unprecedented tools and environments of non-physical interaction," lending urgency to his plea. There may be gray areas around these practices that will be difficult to properly categorize. But we cannot and should not continue to "deny, neglect, misinterpret or trivialize psychological torture" as something distinct from the infliction of "physical pain or suffering." Melzer argues that while modern technology may make a person suffer less physical pain or fewer lasting scars than traditional forms of torture, in light of "international obligations in relation to the prohibition of torture in good faith (Vienna Convention on the

Law of Treaties, arts. 26 and 31) and . . . the evolving values of democratic societies (A/HRC/22/53, para. 14), it seems wrong to exclude the profound disruption of a person's mental identity, capacity, or autonomy only because the victim's subjective experience or recollection of 'mental suffering' has been pharmaceutically, hypnotically or otherwise manipulated or suppressed."[80]

Several treaty provisions, including Article 7 of the International Covenant on Civil and Political Rights, expressly prohibit "medical or scientific experimentation without free consent." And we could, and perhaps should, decide that certain kinds of experimentation constitute psychological torture, such as making a person an unwitting participant in mind control experiments, or using electromagnetic weapons to experiment with disabling a person's mind. Even if those interventions do not cause severe pain. Article 2 of the Inter-American Convention to Prevent and Punish Torture, a regional convention, paves the way, with a definition of "torture" that includes "methods intended to obliterate the personality of the victim or to diminish his physical or mental capacities, even if they do not cause physical pain or mental anguish."

Just as our understanding of self-determination and freedom of thought must be updated to consider emerging technologies, so must our inalienable right to be free from psychological torture. Weapons directed at the mind rob us of control over our brains and mental experiences. This form of manipulation is a stark violation of our freedom of thought. When they are intended to obliterate our personalities, identities, and mental functioning, they should fit within an updated definition of psychological torture as well.

9

Beyond Human

Deprive the brain of oxygen, and within ten or fifteen minutes, its cells begin to break down, leading to irreparable brain damage and death. At least, that's how it's been for most of human history. But recently, a group of Yale University scientists developed a system called BrainEx, which uses an innovative series of pumps and synthetic blood to restore some cellular function to pig brains as long as four hours after the animals were slaughtered for food, bringing us that much closer to the transhumanist vision of super-longevity and Humanity 2.0.

I learned of the Yale scientists' breakthrough in 2016, during an ethics consultation with our Neuroethics Working Group of the NIH Brain Initiative. Nenad Sestan, the leader of the research team, and three of his colleagues met with us to discuss the ethical implications of their research, which was later published in the journal *Nature* in April 2019, with a coveted cover appearance.[1] *Nature* and *Science* are the premier peer-reviewed scientific journals, and they tend to publish only the most notable scientific advances.

What was the Yale team's breakthrough?

Just as a heart or lung bypass can keep a patient alive long enough to repair otherwise fatal damage to those organs, BrainEx brought at least some brain repair closer to reality. By resupplying the dead pigs' brains with oxygen, nutrients, and cytoprotective chemicals present in their synthetic blood analogue, they had restored circulation to the major arteries and small blood vessels along with certain key metabolic functions in the brains. The pigs' brains even showed some synaptic brain activity—activity in the regions where neurons communicate with one another—when a piece of the brain tissue was removed and washed of the perfusate.[2]

While those particular pig brains did not recover any EEG activity—nor any signs of consciousness—that doesn't mean that EEG activity could not one day be restored, just that it wasn't this time. The researchers had included chemical agents in the perfusate (the synthetic blood compound they concocted) that prevented the neurons from firing.

In the short term, BrainEx will allow us to learn more about neurological diseases and functioning. But one day, it may allow us to repair the damage caused by strokes and other traumas. It's hard not to get carried away and hope that someday human brains could be reanimated. All of this makes the goals of transhumanism—the idea that we humans may someday transcend our physical and mental limits, including death—seem within reach.

Not everyone is ready to embrace this Promethean future. A societal movement known as bioconservatism stands in opposition to these lofty goals. To be sure (and as Mary Shelley recognized over two centuries ago) transhumanists' aspirations pose real risks for individuals and society alike. Should they be realized, it's altogether unclear what it will even mean to be human: Will we continue to communicate with each other through spoken and written word or move toward telepathic communication? Will we acquire new abilities or even ways of seeing the world, and how will that change our experience and understanding of others? Will we upload our brains and memories to robotic or synthetic bodies? Who and how should we decide the answer to these questions?

The transformation of humanity has already begun. But most people aren't part of the conversation about whether it is a good thing or not. To realize the promise of neurotechnology and the contours of cognitive liberty, we all must join the debate. One that has been raging for a while now between transhumanists and their opposition.

Humanity 2.0

Transhumanism is a cultural movement aimed at solving the "tragedies" of the human condition—namely aging, our physical and psychological limitations, and suffering.

It draws from across medical and scientific disciplines, including neurotechnology, biotechnology, information technology, and platform technologies such as artificial intelligence, machine learning, automation, and even cryogenic freezing. Transhumanists believe that humans are in transition to the next phase of human evolution, and that radical technological interventions are needed to complete the transformation.[3] According to the philosopher Elise Bohan, author of *Future Superhuman*, the project of transhumanism has already begun, and it will transform our lives, bodies, and minds, ensuring the survival of our species.[4]

As Bohan sees it, in the future, babies will have their entire genomes mapped at birth to predict their risks of disease and develop precision treatments for them. Artificial intelligence will emerge as the most powerful entities on the planet, and humans will have to upskill to compete; human consciousness will be digitized and uploaded, giving us new pathways to immortality; and the experience of work will become radically different from what we now know. She and other transhumanists see death as "the loss of everything that matters."[5] If it were possible for each of us to enjoy robust health indefinitely, humanity would accumulate much more experience and knowledge. "The things that our species could do with that!" she exclaims. "The mysteries of the universe that we could unlock. The problems we could solve. And the depths of each other's' souls that we could explore."[6]

The New Yorker described philosopher Nick Bostrom as "arguably the leading transhumanist philosopher today, a position achieved by bringing order to ideas that might otherwise never have survived outside the half-crazy Internet ecosystem where they formed."[7] Bostrom's essay "A History of Transhumanist Thought" traces the origins of the term to the biologist Julian Huxley, in his book *New Bottles for New Wine* (1957): "The human species can, if it wishes, transcend itself—not just sporadically. . . . But in its entirety as humanity. We need a name for this new belief. Perhaps *transhumanism* will serve: man remaining man but transcending himself, by realizing new possibilities of and for his human nature."[8]

As the founder of the World Transhumanist Association, Bostrom has built on Huxley's ideas, believing as he does that modern philosophers should "acquire the knowledge of a polymath, then use it to help guide humanity to its next phase of existence"—something he thinks he is doing by helping us to understand transhumanism and the "existential risks" that artificial intelligence poses to humanity.[9]

In the short term, Bostrom points to the ways that AI is already displacing humans from jobs, changing the nature of war to automated and remote attacks, enabling greater social manipulation, expanding surveillance capabilities, and making it easier to mislead humans with "fake" videos of real people. But the deepest threat, he argues, "is the longer-term problem of introducing something radical that's super intelligent and failing to align it with human values and intentions,"[10] something he believes transhumanism can help to address. One way to even the playing field is to extend human longevity and enhance our cognitive abilities by uploading our brains to computers—expanding the concept of human dignity to include post-humans within the definition of humanity.[11] The ultimate goal is to live longer, have unlimited intelligence, and live without suffering.[12]

Biopolitics—a new dimension of political opinion—aligns the odd bedfellows of traditional conservatives and extreme liberals against it as bioconservatives. Bostrom believes it could be possible to stake out a middle ground but argues that cultural conservatives have "gravitated towards transhumanism's opposite" instead, embracing a view that stands in opposition to using technology to expand human capabilities or modify human beings.[13]

Bioconservative theorists worry about what transhumanism will mean for human identity and meaning. Philosopher Leon Kass believes that trying to assert technological mastery over humanity could have a dehumanizing effect on humanity and strip meaning from everything in life from sex as an intentional act for intimacy and reproduction, eating for pleasure and not just for honing our bodies, and the dignity of work as artificial intelligence displace human workers.

Francis Fukuyama, another widely recognized bioconservative, has described transhumanism as "the world's most dangerous idea."[14] Fukuyama believes that "liberal democracy depends on the fact that all humans share an undefined 'Factor X' which grounds their dignity and rights"—and he worries that the radical technological enhancement of humans may "destroy Factor X."[15]

In his 2009 book *The Case Against Perfection*, Harvard philosopher Michael Sandel expresses concerns about the impact of transhumanism on human values. What will happen to unconditional parental love when everyone is focused on creating a more "perfect" child? He sees attempts to harness nature as vain endeavors that undermine our humility and solidarity with one another and places the responsibility for our existing limitations on each of us rather than accepting the variation between humans.[16]

But as Bostrom argues, the two groups share some commonalities. Both bioconservatives and transhumanists worry about "existential risks" to humanity. Both agree that technology is likely to change humanity in this century, and that we must grapple with the ethical implications of that transformation. Though bioconservatives favor using technologies just for therapeutic purposes to treat existing medical conditions, while transhumanists believe we are morally obliged to use technology to do that and more to improve the human condition, both contingents worry about their side effects and unintended consequences.[17]

Silicon Valley, not surprisingly, has aligned itself with transhumanism. Google was an early investor in the biotech start-up Calico, which hopes to "answer the most challenging biological questions of our time."[18] The billionaire and venture capitalist Peter Thiel has invested heavily in parabiosis—blood transfusions from younger, healthier individuals to ward off aging—and has even signed up with the cryogenics company Alcor to be deep-frozen when he dies.[19]

While neurotechnology is just one cog in the transhumanists' expansive agenda, how society approaches those advances may very well decide the fate of Humanity 1.0. Between BrainEx and advances in memory decoding, deep brain stimulation, and brain-to-robotic interactions, neurotechnology has already expanded human abilities well beyond what we once thought was possible, and it stands to do even more to minimize human suffering and maximize our well-being.

How should we evaluate the potential upside of these developments against their hypothetical risks? We begin where any good ethical analysis must: by understanding the facts about what is already possible.

The Infinite Mind

Cryonics makes use of extremely low temperatures to preserve bodies until technologies are developed that enable them to be revived or have their brains uploaded into computers. BrainEx may set us on the path to doing so.

Alcor Life Extension Foundation calls it "paus[ing] the dying process." One hundred ninety customers have plunked down between $80,000 and $200,000 for brain- or whole-body-preservation services, including Nick Bostrom.[20] The moment he dies, Alcor will take custody of his body and maintain it in a giant steel bottle flooded with liquid nitrogen.[21]

Nectome preserves brains, too, but its goals are more ambitious. Its website describes its focus on understanding and preserving the connections in the brain and the memories it stores. While in the near term, Nectome expects to provide society with better models to advance the study of the brain, in the long run, it aims to turn those preserved brains into computer simulations that could allow a reboot of the "person" in digital form.

All of this may sound like science fiction, but Antonio Regalado of the *MIT Technology Review,* who has his finger on the pulse of emerging technologies, has advised us to "pay attention to Nectome. The company has won a large federal grant and is collaborating with Edward Boyden, a top neuroscientist at MIT, and its technique just claimed an $80,000 science prize for [cryo]preserving a pig's brain so well that every synapse inside it could be seen with an electron microscope."[22] They have successfully used aldehyde-stabilized cryopreservation on a human brain as well.

While some, like neuroscientist Michael Hendricks, have decried brain and body storage as offering "abjectly false hope," calling "burdening future generations with our brain banks . . . just comically arrogant," Edward Boyden disagrees: "As long as they are up-front about what we do know and what we don't know," he says, he isn't particularly concerned about the ethical implications.[23]

Does self-determination include the right to decide whether to cryopreserve one's brain or body? Does the burden that places on future societies offer a sufficiently strong countervailing societal interest to justify government limitations on individual choice for doing so? People are doing this already. Is it too late to pull the plug on the cryopreserved?

Of course, if it's just immortal consciousness that transhumanists seek, we may soon have other ways to successfully realize their hopes. As implanted brain–computer interfaces become a reality, so may the possibility that those devices could create a living copy of the brain that just might achieve the same goal.

Tracing Our Memories

"I want a beer," the thirty-six-year-old man said, and then asked to have the rock band Tool played "loudly." Those around him heard his words clearly, but he hadn't said them out loud. That's because he was in the late stages of amyotrophic lateral sclerosis (ALS)—a progressive neurodegenerative disorder that had robbed him of any control over his muscles. Before losing voluntary movements, he agreed to have electrode arrays implanted in his brain with the hope that in time they could help him communicate. Three years later and several months after he could no longer even move his eyes, he successfully used his brain–computer interface (BCI) device. "I love my cool son," he told his four-year-old, composing his thoughts at a rate of one character per minute. He then asked for a head massage, and for specific foods to be fed to him through his feeding tubes.[24] "Ours is the first study to achieve communication by someone who has no remaining voluntary movement," said Dr. Jonas Zimmerman, one of the scientists behind this extraordinary feat.[25]

The company that has likely made the most progress toward using BCI to help paralyzed patients like this one is Blackrock Neurotech, whose ethics advisory board I recently joined. Its FDA-cleared NeuroPort Array has been in use for more than a decade. The microelectrode array is implanted in the motor cortex, the area of the brain that naturally controls movement. When patients imagine moving their arm or hand, the device detects those signals and transmits them to an external device, which communicates with prosthetic limbs or other technology, allowing patients like Aaron Ulland, who suffered a major stroke that left him partially paralyzed at just thirty-nine years of age, to move his arm again through an attached brace.[26]

Cognixion is testing its wearable Cognixion One EEG headsets with augmented reality (AR) visors for patients to do the same.[27] The ease and safety of a wearable device like this one will make it the BCI of choice in the near term. But if Elon Musk has his way, implanted BCI may soon catch up.

Musk is best known as the CEO of Tesla and SpaceX, but in the long run, he may be remembered most for another venture. Since 2016, he has been quietly developing futuristic BCI technologies at Neuralink, a company he cofounded with Max Hodak. Neuralink is developing an implantable BCI called "the Link." About the size of a small coin, the Link is designed to replace a tiny piece of the skull. Electrical wires about one-twentieth the thickness of hair and tipped with 1,024 electrodes, extend from the device into the brain, where they pick up electrical signals and transmit them to an external computer or device. (Someday it might manipulate those signals as well.) The Link's battery lasts all day and can be recharged wirelessly. Neuralink is building a robot that can carry out the most challenging aspects of the surgery, with the hope that the procedure will ultimately take less than an hour and not require general anesthesia. [28]

While Neuralink expects that the Link's first clinical trials will be in patients with spinal cord injuries, Musk's ambitions are much grander. Describing it as a "Fitbit in your skull with tiny wires" he hopes to make it "sufficiently safe and powerful that the general population would want" to use it. Musk imagines it augmenting human abilities by transmitting information into the brain (like *The Matrix* but without the painful jack into the back of the head) and allowing us to record and play back our memories.[29] Ultimately, he imagines it transforming humanity, giving us the competitive edge we will need to hold our own against the "existential risks" of artificial intelligence.

Neuralink's video updates—including one of Pager the macaque monkey playing video pong with his mind in 2021—show a steady march of progress. While prominent neuroscientists are quick to point out that other BCI companies are farther ahead, given the relatively easy way the Link will be implanted, its wireless capabilities, and the sheer force that is Elon Musk, it may very well outstrip its competitors, bringing us that much closer to the transhumanists' goal of BCI for everyone.[30] And the transhumanists' vision of uploading our minds.

The Immortal Mind

A recent study at BrainGate—an interdisciplinary multi-institutional effort to develop technology to restore independence to individuals who suffer from brain disorders—suggests BCI might be the tech that will make uploading our brains a reality. In a notable experiment, a paralyzed patient played a game in which he watched colored lights flash on a computer screen and reproduced their patterns by moving a cursor on the screen. The BrainGate BCI device he was wearing recorded the way his neurons fired as he thought about moving the cursor to replicate the pattern. That night, as he slept, the researchers saw that same pattern recurring again and again in his brain.[31] While they were excited about what this teaches us about memory consolidation during sleep, a transhumanist would see the potential for brain recording and, potentially, uploading.

The reality of mind uploading is more complex than its theory: the basic idea is to copy the entire structure and functions of a human brain, including all memory traces, creating a software model that when run on the appropriate hardware would behave like the original, advancing the theoretical possibility of living forever, albeit in digital form.[32]

To realize that vision, we will need safer and less complicated ways of recording brain activity. And at least one company has already beat Neuralink to clinical trials that can do so. Neuralink's cofounder Max Hodak left the company in April 2021 and invested in a rival BCI company, Synchron,[33] that has invented a new and potentially safer, easier, and more scalable way of getting into the brain. Instead of drilling a hole in the skull or strapping a device onto the body, Synchron uses the "stentrode"—a device that looks like a small tube of wire mesh, and remarkably can be implanted via a catheter, much like the stents that physicians use to treat heart patients. The stentrode is fed into the jugular vein in the neck and threaded through a blood vessel that enters the brain. The device is tuned to detect the electrical signals that travel from the brain to give instructions to the limbs and fingers to move. Those signals, relayed through Bluetooth to a device outside the body, are translated by algorithms into computer commands. CEO Thomas Oxley describes it as "bringing electronics into the brain without the need for open-brain surgery."

Four Australian patients with neurodegenerative disorders have been implanted with the stentrode and are able to email, text, and even shop for groceries using only their minds.[34] Synchron has also started clinical trials in the United States. Once widescale safety and efficacy have been established, it's not hard to imagine that even a healthy individual might want a stentrode to more seamlessly interface with technology or reach just a little closer to digital immortality.

Beyond Human Senses

While companies developing implantable BCI devices are focused on therapeutic applications, it's useful to remember that many modern consumer technologies—such as speech-to-text and text-to-speech—were originally designed as accessibility technology. Even eye-tracking in AR and virtual reality (VR) glasses started as accessibility tech.[35]

Transhumanists believe technology can and should be used to enable human beings to transcend our physical and mental limitations. From communicating directly from our brains, to acquiring new senses or new ways of interacting with the world, transhumanists see neurotechnology as a gateway to human evolution.

BCI that is designed to restore our senses may soon be used to expand our senses too. Take Second Sight's legacy product, Argus II, which was surgically implanted into the brains of more than 350 people with retinitis pigmentosa, who had little to no sight. The technology afforded them a novel kind of artificial vision by stimulating their retinas to perceive light and low-resolution images. Its next-generation product, the Orion Visual Cortical Prosthesis System, hopes to bring artificial vision to people who are blind from a wide range of causes.[36] But some people have already enhanced their vision beyond what these devices—or the healthiest eyes—could hope to see.

One of them is the artist Neil Harbisson, who was born with achromatopsia—complete color blindness. Over the last thirteen years, Harbisson has been able to "hear" visible and invisible wavelengths of light through the antenna-like sensor he had implanted in his head. The antenna translates wavelengths of light into vibrations on his skull. As he told *National Geographic*, "I wanted to create a new organ for seeing. . . . At first I could just sense the visual spectrum of light, but I've upgraded it to include the infrared and ultraviolet spectra."

Much as Bohan believes transhumanism will enable us to explore the depths of each other's souls, Harbisson believes his expanded senses have transformed his "understanding of the world" and made it "more profound." Sense enhancement, he argues, is "just the beginning of a renaissance for our species."[37] A renaissance that may soon include even telepathy.

Augmenting Human Communication

Something that has stayed with me since first hearing the CTRL-labs presentation by Josh Duyan in 2018 was the transhumanist ideal he described of moving humanity from efficient "input devices" to efficient "output devices" by enabling us to operate octopus-like tentacles with our minds. If we can operate tentacles, we can communicate brain-to-brain, or even operate swarms of drones—something researchers are already making progress toward. The only limit on what we can accomplish with BCI is our imagination.

A recent experiment shows that brain-to-brain communication is already possible.[38] Researchers at the University of Washington ran an experiment where three different people in three different rooms played a *Tetris*-like game with one another, communicating using BCI. The object of the *Tetris*-like game was to nudge a falling block into the correct position so it could come to rest on a line at the bottom of the screen.[39] Three players sat in different rooms, equipped with EEG headsets and one with a transcranial magnetic stimulation device (TMS), as well. Two of them—the "Senders"—could see both the block and the line but couldn't control the game. The "Receiver" could see and move the block with his mind, but he could not see the line. Each Sender decided whether the block needed to be rotated and would pass that information along using their brain signals via TMS to the brain of the Receiver, who then decided whether to rotate the block. "To deliver the message to the Receiver, we used a cable that ends with a wand that looks like a tiny racket behind the Receiver's head," explained researcher Andrea Stocco, an associate professor of psychology at the University of Washington. "This coil stimulates the part of the brain that translates signals from the eyes. . . . We essentially 'tricked' the neurons in the back of the brain to spread the message that they have received signals from the eyes. The participants have the sensation that bright arcs or objects suddenly appear in front of their eyes." The Receiver used his EEG device to rotate the block in response to the Senders' advice. Five groups of participants played sixteen rounds each, "winning" 81 percent of the time, or thirteen out of sixteen trials, which is well above chance.[40]

Brain-to-text messaging is also on the horizon. In a collaboration comprising researchers from Singapore, China, and the UK, two people communicated whole words to each other using the power of thought. This system—which the researchers call an electromagnetic brain–computer metasurface—uses the same P300 "recognition memory" we saw in chapter 3. The participants successfully sent such phrases to each other as HELLO WORLD, HI, SEU, and BCI METASURFACE during a recent trial. And far more quickly than the ALS patient could order his beer! They were able to send twelve characters per minute—a rate likely to improve substantially when predictive spelling algorithms are also used.[41]

Panagiotis Artemiadis, a professor of mechanical and aerospace engineering at Arizona State University, described a BCI future in which we can "extract information from the brain" to go well beyond our ordinary capabilities, "to control systems and machines that do not resemble anything the brain was made to control." He demonstrated this recently when he tested the ability of a single person to simultaneously control three drones with their mind. When they thought about shrinking or expanding the shape of the drones' formation, an EEG device captured the intention, translated it via algorithms into control commands, and sent them to the drones.[42]

In the past, we made devices that overcame human limitations, using prosthetic limbs, hearing devices, eyeglasses, dentures, and even GPS technology for people like me, who can rarely find their way around otherwise. With BCI and other technology, we may augment our senses to see spectra of light we have never seen, hear soundwaves we could not hear, even smell danger that we could not sense before. Or even to eliminate certain human experiences that until now were fundamental to the human condition—like suffering after loss or tragedy.

The End of Human Suffering?

If your brain had a switch to turn off suffering, would you use it? I wonder if I would have to lessen my suffering after our daughter Calista's passing, regaining the year that followed which remains just a hazy blur in my mind. That may now be possible with a brain implant that disrupts the signals that lead to pain and suffering.

A team at the University of California, San Francisco, implanted a BCI device in Sarah, a patient who previously suffered from intractable depression. "When we turned this treatment on, our patient's depression symptoms dissolved, and in a remarkably small time she went into remission," Dr. Katherine Scangos, a neuroscientist and psychiatrist told CNN. "It was like a switch."[43] A year later, Sarah is free of both depression and any side effects from the treatment.

The study team began by mapping the areas of Sarah's brain that became active when she was experiencing the worst symptoms of her depression. Then they implanted a small wire into her brain that delivered a pulse of electricity to disrupt the signals. While the treatment is highly individualized and will likely require years of refinement, it offers great promise. Sarah described "laugh[ing] out loud," and the "joyous feeling [that] washed over" her the first time the team used stimulation.[44] Could we map other forms of suffering and bathe future humans' brains with that same joy? Should we?

Flow Neuroscience, which purchased the company Halo Sport, offers a wearable device to alter the brain with transcranial direct current stimulation. Olympic athletes used it to stimulate their motor cortexes and improve their rate of learning in preparation for the 2016 Summer Games in Rio de Janeiro.[45] That same technology is now being used to alleviate the symptoms of depression. They claim a success rate of 83 percent.[46]

Treating depression isn't the same as eliminating suffering; Sarah can still experience joy and sadness, and even more so now that her depression is under control. But therapeutic applications to disrupt the symptoms of depression point the way for transhumanists to realize their long-held goal of eliminating human suffering. It sounds appealing, but it may have unintended consequences.

Writer Joelle Renstrom derides the transhumanists' call to end human suffering as "irresponsible, socially divisive, and inherently egotistical in its assumption that suffering is universally undesirable and meritless."[47] Physical pain, after all, alerts us to impending danger. Transcending sadness and tragedy, she argues, helps us learn how to thrive in the face of adversity and develop the confidence that we can prevail when we face future challenges. Moreover, it is the duality of sadness and happiness that makes "happiness more powerful," by giving us the full breadth and depth of emotional experience. Instead of maximizing our well-being, pursuing the end of suffering may "beget the very suffering it seeks to abolish."[48]

Bending the World to Our Will

As neurotechnology, artificial intelligence, and robotics continue to advance, we can expect a future that bends more to our will and adapts more to our minds. When neural interfaces become the predominant way that we interact with our technology, we will be that much closer to a future in which policies are evaluated based on their impact on our "collective cognitive capital," as Emily Murphy, of the University of California, San Francisco College of Law, put it. It's a concept she coined to describe a new approach to public policy that would evaluate how programs or policies impact our brain health and functioning as a collective asset for humanity to maximize.[49]

Penn State researchers are working to create a future in which human workers wear EEG headsets to provide input to their robot coworkers, who calibrate the pace of work to our state of mind.[50] In one experiment, research participants donned EEG headsets that detected signs of stress and increased or decreased cognitive load. Their robotic coworkers reacted by slowing down, speeding up, or keeping a steady pace, giving the worker just the right amount of room to maximize their productivity without stressing them out.[51]

Microsoft's Human Factors team similarly helped Microsoft adapt its products to be more responsive to our brain health and functioning. when working remotely. Based on information captured by EEG headsets, they discovered that during video calls our brains struggle to keep track of everything from misaligned eye gazes to missing body language and hand gesture, tiring us out after thirty to forty minutes and raising our stress levels for another two hours. When given the chance to rest between meetings, the brain resets and our stress levels decline, allowing us to become more focused and engaged in subsequent meetings. Based on this knowledge, Microsoft changed how it schedules meetings, and even redesigned its popular Outlook calendaring program to enable other companies to start meetings five minutes after the hour or half hour, to allow for breaks.[52]

Even our experience with art is starting to bend to our brains. The start-up company Cuseum hopes to adapt our future art museum experiences to maximize our brain engagement. In a ten-month study, it measured museumgoers' brain wave activity as they encountered real paintings, 2D images of the same works on an iPad, and the works seen through an Oculus virtual reality headset paired with an iPhone. Based on the participants' brain activity, it found that their immersion in the art was just as strong when seen in AR/VR as when viewed in physical space—if not stronger. While we should take these findings with a grain of salt—as their research was based on a small sample size, and the company's goal is to sell technology to art museums—it nevertheless points to a probable future in which our experiences with work, art, and technology may constantly adapt to optimize our brain activity.[53]

Artists from the American cellist YoYo Ma to Korean performance artist Lisa Park have begun using neurotechnology as part of their artistic ventures, as well. After a presentation I gave at the World Economic Forum meeting in Davos in 2016, YoYo Ma approached me to ask whether there are any good studies on using consumer neurotechnology while listening to music. Two years later, he surprised unsuspecting Montrealers with a free concert at the Place-des-Arts Metro station in downtown Montreal, while joined by Olivier Ouillier of Emotiv to demonstrate the impact of music on the brain.[54]

Park's project, Eunoia, which means "beautiful thinking," "manifests" her brain waves into sounds.[55] Masaki Batoh documented the devastation of the Great East Japan earthquake and tsunami by using EEG headsets to measure the brain waves of victims and play them back as music for his album *Brain Pulse Music.*[56]

The progression toward Humanity 2.0 is already underway; the ground is shifting beneath us as our environment starts to become increasingly responsive to our brain activity.

But while that transformation has already started, a public dialogue has not yet begun. If we want to balance our cognitive liberty with societal interests, it's time for everyone to join the conversation.

A Path Forward

In May 2010, a team of scientists from the J. Craig Venter Institute announced that they had created the world's first self-replicating synthetic cell—an organism whose "parents" were the chemicals that are the building blocks of nature and a computer program. The media went wild with the "horrifying and wonderful" idea of "whipping up a new life form in a lab."[57]

Soon after, President Barack Obama asked our Bioethics Commission to make it our first order of business to deliberate about the milestone and provide a set of concrete recommendations about an ethical pathway forward.[58] We approached the task through a process of deliberative democracy—engaging with the existing literature, hearing from and exchanging ideas with scientists and engineers, faith-based and secular ethicists, and the many other stakeholders who would be affected by these advances. We held public meetings and listened and exchanged ideas on different perspectives about the science and ethical approaches to address potential risks and benefits, and the relevant legal and regulatory frameworks worldwide.

This model—of engaging in democratic deliberation via bilateral dialogues—is a conceptual framework that our chair, Dr. Amy Gutmann (then president of the University of Pennsylvania), had refined in her own scholarship. In our first report, *New Directions: The Ethics of Synthetic Biology and Emerging Technologies*, we endorsed "the importance of robust public participation in the development and implementation of specific policies as well as in a broader, ongoing national conversation about science, technology, society and values."[59]

This approach may help bioconservatives and transhumanists find more common ground, even as their competing perspectives inform the careful oversight these developments will require. Policymakers worldwide will need to better engage the public in an ongoing dialogue about neurotechnology. Scientists and entrepreneurs will need to be as transparent as possible about their potential uses and listen to competing views and a broader set of stakeholders.

Just as members of different religious and secular traditions, civil society groups, and other stakeholders must be heard and their perspectives valued, scientists need to become an increasing part of the conversation, making clear what is scientifically possible and impossible. To that end, they should receive better training in public communication. We will need to continue to equip society with scientific and ethical literacy by ensuring that manufacturers provide clear and effective disclosures. Private foundations, public institutions, and governments all have a role to play. As we wrote in our report, "scientific research and public education about science are best approached as mutually related, even mutually dependent, endeavors."[60] Public literacy about neuroscience and its technologies is critical to empower people to exercise their rights to mental privacy, self-determination, and freedom of thought—that cognitive liberty comprises.

As for the risks of the kind of radical human transformation advocated by transhumanists?

My fellow commissioner, the late philosopher John Arras of the University of Virginia, asked: "What is the proper attitude in dealing with events which are in low probability range, but high in impact? . . . One way of handling risk is to be proactive: Go full speed ahead, worry about risks later. This supports the values of scientific freedom and focuses on the benefits that derive from it. The other extreme—the cautionary proposal . . . there must be a promise of mitigation before one goes forward with research." Arras advocated for an Aristotelian approach, which was to take the middle road: *consider all views on a subject and then present findings based on deduction and practical consideration.* A view we as a commission embraced. "We call it prudent vigilance," Arras explained. "We recommend having ongoing assessment as the risk develops. We argue that research goes forward but with lots of safeguards."[61]

Ongoing assessment of the risks of neurotechnology is already happening. The Organization for Economic Co-operation and Development (OECD), the United Nations, the Council of Europe, the International Neuroethics Society, the IEEE Brain initiative, and the NeuroRights Foundation, to name but a few, have held convenings to discuss ethical progress in neurotechnology. A group of corporate executives and scholars have called for a White House task force to "craft a roadmap for the effective governance of applied neuroscience technologies." Our Bioethics Commission launched a similar dialogue with our *Grey Matters* reports, but there has not been another US Presidential Commission on Bioethics since 2017, with the necessary convening power to continue the conversations.

Other disruptive technologies have developed effective models and best practices that can serve as a roadmap. CRISPR, a technology that allows scientists to make precision edits to our DNA, has garnered international attention and dialogue to develop ethical norms to maximize its benefits for society. Stanford Law professor Henry T. Greely has chronicled the history of those discussions.[62]

In January 2015, a small working group of scientists, ethicists, law professors, and more came together to focus on the most ethically concerning aspect of the use of CRISPR in humans—editing embryos, which could impact future generations by making changes to our "germline" DNA—the DNA we pass on to future generations through egg and sperm cells. The group quickly agreed that CRISPR was an important advance for humanity, but that it should not yet be used for human-germline editing. They published an editorial discouraging that use and calling for an international forum to engage a broader set of global stakeholders on its applications.[63]

Soon thereafter, the National Academy of Science and National Academies of Sciences, Engineering, and Medicine created a Human Genome Initiative to guide future conversations. The initiative convened an International Summit on Human Gene Editing in late 2015, which was jointly sponsored by the Royal Society in the UK and the Chinese Academy of Sciences. At the end of the meeting, the organizing committee issued a statement on their own behalf that "It would be irresponsible to proceed with any clinical use of germline editing unless and until . . . the relevant safety and efficacy issues have been resolved." They called for ongoing conversations to establish "broad societal consensus" about future applications of CRISPR technology.[64]

To achieve that broad consensus, they identified the important role of government investment and participation in supporting an international forum to engage representatives from the sciences, medicine, public policy, patient and patient advocacy groups, industry, and more. The Human Genome Initiative remained at the helm to bring those conversations to fruition, ultimately leading to a consensus report that was issued as the "'official position' of the Academies" in 2017.[65]

The path to minimizing the risks of the technology hasn't been without bumps. A bombshell revelation at an International Summit on Human Genome Editing in Hong Kong in November 2018 was a stark reminder about the importance of ongoing prudent vigilance. A researcher from Shenzhen revealed during the meeting that he had used CRISPR to edit the genomes of embryos to attempt to make them resistant to HIV. And more shockingly, he had transferred those embryos to a woman's uterus who fell pregnant and delivered genetically edited twins. The summit organizing committee, world scientific academies, and prominent scientific leaders decried the research as "deeply disturbing" and "irresponsible."

We could see this as a failing of international norms to police unethical uses of technology. But I think it teaches us otherwise. Because his actions ran counter to established norms, the global condemnation was swift and unambiguous. This has prevented other scientists from following in his footsteps. His grave misstep also reminds us about the interrelationship between norms and their enforcement. Prudent vigilance requires regular oversight of emerging technologies, and the adoption of enforceable laws and regulations to address risks as they emerge.

Since the November 2018 meeting, the various scientific academies across the world have established a commission to "develop a framework for scientists, clinicians, and regulatory authorities on the appropriate use of human germline genome editing."[66] A group that continues to monitor the technology as research evolves, updating international norms, engaging stakeholders in an ongoing dialogue, and providing governance recommendations to guide regulators worldwide.

The time is now right for us to establish an international body to provide the same kind of oversight for neurotechnology. To engage society in an ongoing process of democratic deliberation, and to maintain prudent vigilance over the progress of neurotechnology. That body should include delegates from scientific organizations worldwide, representatives from ethics, religious, and civic organizations, members of the public, patients and their families, and other stakeholders from across society. We will also need the focus and determination of every country to strike the right balance between progress in neurotechnology and individual liberties, and to enforce international human rights at a national and local level.

We can't yet know all the risks and benefits that will come from the ongoing project of transhumanism. But we can continuously assess both the scientific progress and the risks it entails to maximize the benefits of neurotechnology for humanity.

10

On Cognitive Liberty

When Shoshana Zuboff coined the concept of surveillance capitalism, our personal data had already been widely commodified and our ability to claw it back largely gone. With neurotechnology, it's not too late to protect against that same fate for our brains. We stand at a fork in the road—where the coming dawn of neurotechnology could change our lives for the better or lead us to a more dystopian future where even our brains are hacked and tracked. The choice is still ours to make.

How do we choose the right path? By recognizing a new human right to cognitive liberty. Legal scholars Brandon Garrett, Laurence Helfer (also elected US representative to the UN Human Rights Committee), and Jayne Huckerby showed the way in a recent article proposing a new international human right to claim innocence after being convicted of a crime. Recognizing such a right, they argued, would have significant "symbolic, strategic, normative, and enforcement benefits."[1] It would have symbolic value by making a clear global norm; it would strategically empower social groups and mobilize them to further define its contours; it would change the default rules in favor of claims of actual innocence worldwide; and bring the right within an existing legal framework with enforcement mechanisms.[2] Their approach shows why it's important that we recognize a new human right to cognitive liberty, as well—and how doing so will enable us to update the human rights it impacts—the rights to privacy, freedom of thought, and self-determination over our brains and mental experiences.

Recognizing an international human right to cognitive liberty would make it a clear legal priority to protect our mental experiences as much as our other physical ones. Doing so would guide future conversations about the implementation of neurotechnology—whether used in healthcare, education, in the workplace, or by the military.[3] As neurotechnology advances and governments and corporations can increasingly invade our brains, what we are feeling, and what we are thinking, this symbolic benefit can't be overstated. John Stuart Mill argued in *On Liberty* that the "appropriate region of human liberty" comprised "the inward domain of consciousness." But the world he lived in was not a world in which even our minds could be revealed. Now, our last fortress of privacy is in jeopardy. And our concept of liberty is in dire need of being updated.

Adopting a right to cognitive liberty would also shine a blinding light on the many reasons we will want to embrace neurotechnology, but also the grave risks of doing so without adequate protections in place for individuals. This would strengthen the growing social movements that are focused on minimizing the risks of neurotechnology by protecting the inner domain. And make clear countries', corporations', and individuals' obligations to respect the cognitive rights of others, while creating a way to hold actors accountable if they violate those obligations.[4]

In some countries, international human rights law trumps any existing or conflicting laws so the right would go into effect immediately. Others would adopt laws to implement the right to cognitive liberty in specific contexts—regulating, for example, whether employers could use neurotechnology in the workplace and placing limitations on the brain wave data they collect. Existing structures and institutions in international law would also help to monitor violations of the right to cognitive liberty.

As Garrett, Helfer, and Huckerby describe it, the process of recognizing a new right involves "'identifying previously unarticulated aspects of old human rights' . . . or articulating 'newly recognized aspects of existing rights.'"[5] A new "General Comment" could be adopted to recognize the right to cognitive liberty and its relationship to the bundle of rights it encompasses—and define and elaborate on how it would apply in specific contexts. The rights it is derived from, and impacts, will also need to be updated to be consistent with a new right to cognitive liberty.

International human rights are meant to evolve over time, as circumstances in the world demand doing so. There are several ways that happens, such as through observations and recommendations written by governing bodies like the Human Rights Committee, judicial opinions in domestic cases, and revisions to existing General Comments that elaborate on how we should understand the meaning and scope of recognized rights.

Explicitly making mental privacy part of our existing right to privacy would shield our brains and mental processes from others in most instances, including identifying information from our brains, automatic processes, memories, and silent utterances in our minds. As a relative right, intrusions upon our mental privacy will be subject to the three-part test that is used in international law—legality, necessity, and proportionality. Is there a law that allows for encroaching on our mental privacy—such as a specific one that allows employers to monitor certain employees for fatigue? Is it necessary to do so to satisfy a compelling social interest like preventing mass fatalities from a trucking accident? Is the impact on the person's rights proportionate to the societal interests being achieved? Are we intruding no more than necessary to protect the significant interests of others?

Freedom of thought, by contrast, is an absolute human right, and we will need to update our understanding of it to apply to contexts outside of religious freedom. Because of that, we must carefully interpret it to apply only to robust thought, like our memories and silent thinking and imagery, so we don't make ordinary human interactions unlawful.

Without an individual right to self-determination, we cannot exercise our other human rights, so we must make explicit what has been an implicit right across the globe. That includes the right to informational self-determination to access information about ourselves, the right to choose how we will change our own brains and mental experiences, and the right to our capacity to choose by making it unlawful to disable or control a person's mind. Although it is a relative right like mental privacy, because it is crucial to exercising our other rights, the tests of necessity and proportionality will rarely be met.

More benefits and risks will emerge as neurotechnology becomes part of our everyday lives and other rights may need to be updated to address them as they emerge. But we shouldn't wait for those to arise before acting or we will miss the opportunity to choose the path forward for society.

We are past due on the urgent need to recognize the right to cognitive liberty over our brains and mental experiences. With prudent vigilance and democratic deliberation, we must continue the process of deciding how that liberty can and should evolve over time.

Afterword

Since *The Battle for Your Brain* first appeared on bookstore shelves, a seismic shift has occurred in the landscape of neurotechnology and AI. The initial writing of this book focused on the potential, the perils, and the foreseeable limitations of neurotechnologies in going from brain state reading to "mind reading." But recent technological developments have brought us considerably closer to the future envisioned in the book, much faster than was thought possible a few years ago. The late 2022 debut of OpenAI's ChatGPT and the subsequent wave of rival large language models from major tech companies have not only expanded the AI narrative, they've fundamentally changed it. These models have enabled researchers to make impressive strides from simply interpreting brain states to beginning to decode the stories we hear and imagine in our minds. These developments undoubtedly informed *Nature Electronics'* decision to name 2023 "The Year of Brain-Computer Interfaces," and prompted the influential journal *Nature* to kick off 2024 by listing brain-computer interface as one of the "7 Technologies to Watch."[1]

The good news part of this story has continued to get better. The recent progress in implanted neurotechnologies has been particularly hopeful, potentially soon empowering people with severe neurological conditions, like speech deficits and paralysis, to reclaim their self-determination. The story of Gert-Jan, a forty-year-old who was paralyzed following a bicycle accident, exemplifies this progress. Scientists developed a "digital bridge"—a brain-computer interface enabling Gert-Jan to convert thoughts into leg movements.[2] His ability to stand and walk again signifies a great technical achievement. But it's more than that: it underscores the promise of the right *to* cognitive liberty as a right to access and use neurotechnologies that can benefit us, and not just a right *from* interference with our mental privacy and freedom of thought. With the FDA granting more breakthrough designations to brain-computer interface companies like Precision Neuroscience, Paradromics, InBrain Neuroelectronics, and Cognixion, we can expect even more good news to come.

The past year has also seen the concept of cognitive liberty become a pivotal part of the global discussion on neurotechnology and AI governance. Since the book's release, my journey has included a whirlwind of dialogues and debates, including talks at the International Association of Privacy Professionals, OECD, UNESCO, TED2024, Aspen Ideas Festival, Google, Meta, Apple, and IBM, as well as with numerous government agencies, and at the World Economic Forum in Davos. In late 2023, 194 member states entrusted UNESCO with drafting recommendations on neurotechnology ethics. Like their global standard on AI ethics adopted by all 193 member states in November 2021, their upcoming recommendations for neurotechnology are expected to be highly influential. Cognitive liberty has already become part of those discussions, though to be sure there remains a tension between the principle and existing business models built on commodifying personal data.

Perhaps the moment that most captured public attention in recent times came on January 29, 2024, when Elon Musk's post on the platform X instantly grabbed global headlines: "The first human received an implant from @Neuralink yesterday and is recovering well. Initial results show promising neuron spike detection."[3] This unconventional announcement set off a flurry of media speculation, suggesting a potential milestone in brain-computer interface technology.

Musk's announcement of Neuralink's first product, Telepathy, kindled hopes for people suffering from substantial limitations in speech and movement to a future where they could control devices and even limbs "just by thinking."[4] Rival companies including Synchron, Precision Neuroscience, and Blackrock Neurotech may have reached this milestone before Neuralink, but the pace at which Neuralink has moved is noteworthy. Only months after gaining regulatory approval, Neuralink has implanted its first device in an adult quadriplegic patient, showcasing its ambition to make its brain-computer interface device a modern reality. While it will take months to years to learn more about the safety and efficacy of their device, Neuralink has made clear its vision extends beyond these immediate medical applications to "unlock human potential" in the future.

A key innovation in Neuralink's approach is the use of a proprietary surgical robot, which became necessary since the ultra-fine threads of their device defy manual insertion by the human hand. It may become the surgical robot more than the device itself that catalyzes the field, where one of the bottlenecks to scaling implanted brain-computer interface devices will be the limited number of neurosurgeons skilled at performing these operations.[5]

These landmark advances in neural interface technology are increasingly converging with advanced AI in interpreting complex neural signals—with generative AI models like those that gave rise to ChatGPT already being used to make significant research strides. It's in this context that the study led by Dr. Alex Huth at University of Texas at Austin, published in April 2023, intersects with these advances. Using GPT-1, the most advanced model available at the time to the researchers, Huth's team embarked on a novel experiment: training an AI decoder to reconstruct narratives from brain activity captured by fMRI. Participants listened to hours of podcast stories while their brain patterns were recorded, providing a rich dataset for the AI to learn from.[6]

The results offered a glimpse into a future that Neuralink and others are now actively pursuing: AI systems that can not only decode individual words and phrases but also piece together overarching narratives from our brain activity. UT Austin's decoder wasn't perfect at doing so—sometimes it only got the gist right, not the words, or it wrongly decoded the thoughts of the trial participants altogether. But this study still represented a significant leap toward AI-powered neurotechnology understanding and translating the complex tapestry of human thought. And the foundational AI models have already improved substantially since then.

To address potential societal concerns, the researchers also ran additional experiments. They tested whether the AI decoder, when trained on one individual's brain activity, could be applied to effectively decipher what a different person whose data was not used to train the system was hearing or imagining. They also tested whether a person could intentionally disrupt the decoder's accuracy by being uncooperative. These results showed the decoder's effectiveness significantly declined when applied to a different person, and that it could easily be defeated by the unwilling person. Yet these findings offered some reassurances against fears of imminent coercive neural interrogation. But as testament to how quickly things are changing thanks to rapid advances in generative AI, those reassurances were quickly dashed by even more recent findings from Dr. Huth's lab: a "converter" they have developed allows their decoder, trained on one participant, to effectively decode a different person's brain activity with only minimal training data by that person.

And yet a more likely scenario for the future involves people voluntarily training decoders on their own brain activity to allow them to interact with devices embedded with brain sensors, such as smartwatches, earbuds, headphones, and immersive technologies. As AI becomes increasingly proficient at interpreting neural signals, we are entering an era where our innermost thoughts and intentions could directly steer our interactions with the digital world.

This vision of the future came into further focus during the Meta Connect conference in September 2023. During a fireside chat away from the limelight, Meta's chief scientist, Michael Abrash, and chief technology officer, Andrew "Boz" Bosworth, discussed the company's innovative electromyography (EMG) brain sensors, set for release in early 2025. These wristwatch-embedded devices reflect major advancements in AI's ability to interpret nerve impulses, promising more intuitive interactions with virtual environments. As Abrash explained, these sensors are designed to "learn you," symbolizing a shift from technology that merely responds to our physical commands to one that learns, adapts, and interacts with our mental states.[7]

In a meeting Meta convened in late 2023 with global ethics experts, I pressed Meta's EMG product team on this conversation, and they in turn portrayed their EMG watch that will first ship as having far more modest capabilities than the far-reaching, exploratory ambitions their research endeavors foretell. This was highlighted in study they published in *Nature Machine Intelligence* in 2023, which marked a significant advancement from using stationary and bulky fMRI technology to decoding perceived speech with far more portable EEG and magnetoencephalography (MEG) technologies. Their sophisticated AI system, including a unique "subject-embedding layer," showed remarkable accuracy in interpreting specific speech segments from brain activity recorded during extensive listening sessions, referred to as "brain conversations." These were essentially sessions where participants listened to stories or sentences while their brain activity was recorded. While still not "mind reading" of unintentionally communicated thoughts or ruminations, the results mark significant progress in AI's ability to interpret human brain activity.[8]

Meta's research, alongside subsequent developments like the Sydney-based DeWave system, represents yet another leap forward reported over the past year in decoding silent speech from portable EEG technology. In a preprint that has so far been met with skepticism about the methodology and findings by others in the field, the researchers describe DeWave as using a cutting-edge method to translate brainwave patterns into text. It's akin to a sophisticated system that acts like a translator, converting the unique language of brain signals into words we can understand, creating a dictionary linking EEG signals to language tokens. DeWave provides a glimpse into a not-so-distant future where AI might interpret brainwaves with ease, potentially as easily as we understand words on a page. At least for now, though, the quality of brain decoding is still far less than that of language translation and speech recognition.[9]

The Apple Vision Pro, launched in February of 2024, which is capable of interpreting eye movements and muscle activity to infer mental intentions, may well be the first widely available product powered by our minds, marking the beginning of a new era in neurotechnology. And with Apple's recently filed patent application for integrating EEG brain sensors into devices like AirPods, the potential for seamlessly integrating brain-sensing capabilities into our daily lives becomes increasingly plausible.[10]

This trend extends beyond tech giants; companies like OpenBCI have already showcased the integration of EEG and EMG sensors distributed across the body in the remarkable TED talk delivered by Conor Russomanno, where he stopped talking about their device, and ceded the stage to Christian Bayerlein, a man with advanced spinal muscular atrophy, who used his brain activity to fly a drone over the live TED2023 audience using hardware and software created by OpenBCI.[11]

The burgeoning brain-computer interface sector, now fueled by over $33 billion in private investment, is also driving a wave of startups entering the space.[12] Companies like Neurable and NextSense started taking preorders for multifunctional devices embedded with brain sensors, while Emotiv launched its MN8 earbuds featuring in-ear technology. This surge in development and investment signals a tipping point in the integration of neurotechnology into our everyday lives.

It's yet to be seen whether these emerging products will prioritize improving mental health or follow a path focused on extracting and commodifying neural data. This pivotal choice will shape whether these technologies align with human flourishing or perpetuate models that exploit our attention and data.

Perhaps it's a hopeful indication of what's to come that on the same day Apple introduced its Apple Pro Vision, it more quietly unveiled a new suite of mental health features including mood tracking, journaling, exercise minutes, sleep data, and a mental health questionnaire.[13] Or that other entrants to the space have also made clear their intention to focus on mental well-being. Neurable's headphones promise new insights on whether you are focused or your mind is wandering, and to make suggestions about when it might be time to take a break and reset. Products like Cala Health's FDA approved kIQ armband, which provides precise neurostimulation for essential tremor and Parkinson's, or the portable tDCS Flow device approved in the UK and EU for home neurostimulation treatment for depression, point to a future of not only greater clarity about what's happening in our brains but new ways to treat them as well.

And yet, Neuralink's ambitions and Meta's and scores of other companies' innovations at the convergence of AI and neurotechnology are not just about technological progress. They are harbingers of a future where the sanctity of our innermost thoughts may become accessible to others, from employers to advertisers, and even government actors. The prospect of machines decoding our thoughts, dreams, and fears is as intriguing as it is unsettling, leading to increasingly more urgent discussions on neurotechnology governance.

This is how we find ourselves at a moment when we must be asking not just what these technologies can do, but what they mean for the unseen, unspoken parts of our existence. Alongside or in addition to the uptick in conversations about the governance of AI has been a growing focus on governing neurotechnologies. From the release of an important UNESCO report[14] and their mandate to quickly develop recommendations on the ethics of neurotechnology, to a major report issued by the Information Commissioner's Office in the UK[15] analyzing the impact of neurotechnologies and neurodata on privacy, to the U.S. and other countries advancing legislation to protect neural data, and countless other convenings, meetings, and publications, responsible governance of neurotechnologies and neural data has gained significant attention.

And yet these expert-level discussions have not yet led to a widespread public awakening akin to a "ChatGPT moment." Instead, the public's first encounters with neurotechnologies continue to be in settings far removed from ethical discussions about the technologies. This may be desensitizing people to the risks and normalizing the technology without having the chance to grapple with its transformational impact on the human experience.

Consider the visually stunning and groundbreaking generative AI artwork exhibit entitled *Unsupervised* by artist Refik Anadol at the Museum of Modern Art in New York, which ran from 2022–2023 (and was since acquired by MoMA).[16] Anadol used machine learning to interpret and reimagine images of artworks in MoMA's collection. A far less-publicized aspect of the exhibit was the collaboration with Neuroelectrics, a neurotechnology company, which asked museumgoers to use wearable neurotechnology to record their brain signals as they gazed at the artwork.[17] Immersed in the dramatic installation, these visitors likely remained oblivious to the quiet surrender of their mental privacy.

There is a clear risk to introducing the public to the novel world of neural interface technology in settings like these or at the perfume counters or in marketing gimmicks described in this book. Novel technologies often become normalized through recreation and play, because the ways we engage with them seem purely fun and exploratory. In these low-stakes settings, we become familiar with the technology and learn positive associations with using it, which can obscure the deeper implications for mental privacy and cognitive liberty. Indeed, the more immersive and entertaining the play experience, the more we develop emotional attachments to technologies.[18]

These examples underscore a fundamental challenge: as we become more familiar with and emotionally attached to neurotechnologies, we may unwittingly acquiesce to a new norm of neural surveillance. Behavioral economists B. Joseph Pine II and James H. Gilmore's concept of the "experience economy" is exemplified by this fusion of technology with personal experiences.[19] Prominent figures like Canadian singer-songwriter Grimes amplify this narrative. When she announced several days after the release of the hardback version of this book that she wanted a consumer brain-computer-interface device[20] for her birthday, Neurosity, the maker of the Crown EEG device, gladly complied, catalyzing a staggering increase in their orders.[21]

Each new encounter with neurotechnology—whether through research, art, or consumption—is subtly adapting us to a future in which neural surveillance will become a mainstay of society. As the scholars Judy Rhee and Evan Selinger warn in their essay "Normalizing Surveillance," the harms from surveillance might emerge long after the fact.[22] One stark illustration is the 2018 complaint filed by Canadian actor Jennifer Kobelt against the now-defunct NXIVM cult, claiming that a NXIVM physician had subjected her and at least eighty other people to neural surveillance with EEG headsets while they watched graphic videos as part of a "fright study."[23] The real purpose of this research was allegedly to identify the most impressionable women for induction into sexual slavery.[24]

The broader public needs to become key stakeholders in ongoing conversations about neurotechnologies. The chilling effect of government surveillance on free expression is well documented, and the prospect of government neural surveillance raises even graver concerns for thinking freely.[25] Recent reports[26] of Chinese researchers using facial expressions and EEG for ideological education, or to detect a "brainwaves spikes" from people watching prohibited pornographic content[27] highlights the urgency of raising public awareness.

Which brings us back to cognitive liberty.[28] This is about more than preventing unwanted mental intrusions; it's a guiding principle for human flourishing on the road ahead. We should move quickly to affirm broader interpretations of self-determination, privacy, and freedom of thought as core components of cognitive liberty. Important headway is being made toward this goal worldwide including a global first in Chile of a legal opinion finding an individual right against the commercial collection of brain data.[29] As states and countries move forward with soft and hard law governing neurotechnologies and the data they collect, they should set a mental privacy floor, calling for edge storage and processing of neural data, to keep sensitive data localized and encrypted on personal devices and overwritten unless individuals consciously and affirmatively choose otherwise. We've put strong bulwarks into place around health and financial data; data about our brains and mental states deserves the same level of protection, if not more.

But protecting cognitive liberty is about more than just passing laws and recognizing rights for individuals. It will require a holistic ecosystem change to safeguard human flourishing in the digital age. Cognitive liberty should be at the heart of educational programs aimed at digital literacy and citizenship, a core design principle for commercial design and redesign of products, and a guiding principle in digital policy frameworks. Public awareness campaigns are essential to shifting the demand side of this equation and empowering consumer choices, as are robust digital literacy and citizenship educational programs to foster self-determination in the youngest and most vulnerable in society as well as lifelong learners.

A paradigm shift is needed to align commercial design with cognitive liberty, which will require investments by venture capitalists and consumer demand for products that enhance, rather than diminish, critical thinking and focus. Apple's recent unveiling of cognitive wellness features, emphasizing not just user engagement but genuine self-awareness and mental well-being, underscores the power of platforms to cultivate cognitive liberty. And commitments by companies like OpenBCI to use cognitive liberty as the product design framework set a new standard for other consumer neurotechnology companies.

Just as Lex Fridman's recent podcast interview with Mark Zuckerberg in the Metaverse highlighted the potential for digital platforms to foster intimacy and empathy, there is an opportunity for tech companies to pivot toward fostering meaningful, privacy-respecting interactions.[30] The Senate judiciary hearing on online child safety, held on January 31, 2024, during which Mark Zuckerberg rose from his chair to apologize to the families of abuse victims for the harm caused by his platform[31] could be a turning point, highlighting the need for companies to prioritize mental well-being and ethical data practices. Platforms like pi.ai are taking the lead in doing so, designing their large-language models to prioritize empathetic interaction with users and to foster their critical thinking skills.

To motivate more tech giants to follow this path, we need to recalibrate the incentives driving their business models. Drawing parallels from the renewable energy sector, governments can incentivize companies that prioritize user self-determination, privacy, and ethical data handling through tax breaks, subsidies, and grants. Such incentives can create market dynamics favoring the development of technologies and business practices aligned with cognitive liberty principles, just as incentives have led us to reach the tipping point in twenty-three countries toward the adoption of electric vehicles.[32]

The question now is whether we will passively accept a future where our thoughts may no longer be our own or take decisive action to champion cognitive liberty and shape a future where our neural narrative remains our own. The path we choose will not only define our relationship with technology, but also what it means to be human in the digital age.

Acknowledgments

This book is the culmination of more than a decade of thinking, research, conversations, scholarship, presentations, and dialogue with others. I am grateful to the many people who have been involved and who have given me the support needed to make it possible. Even though they aren't all mentioned here, each of them has had an important impact on my ideas and writing of this book.

A few played an outsized role in making the book happen. My literary agent, James Levine, has been by my side for nearly a decade in the development of this book. He helped to sharpen my ideas and writing and has been remarkably patient and generous with his time and insights. I am fortunate to have worked with my extraordinary and astute editor at St. Martin's Press, Tim Bartlett, who believed in me and the importance of this book, and who has given me crucial feedback on the substance and writing. I owe an enormous debt of gratitude to Arthur Goldwag for his feedback, editorial assistance, ideas and additions to the book, and encouragement along the way. He helped me build my confidence in the writing process and ideas expressed in this book. Toby Lester gave me terrific early feedback about how to tackle this book and my writing of it. And Eliani Torres made numerous suggestions in the copyediting stage that improved the readability and clarity of my writing.

I am also grateful to my many friends and colleagues in neuroethics, law, and philosophy who have supported me with their comments, conversations, and cheerleading along the way. Hank Greely provided endless nudges, wise counsel, and encouragement to "get it done." Nicholas Quinn Rosenkranz helped to sharpen my thinking and offered invaluable insights at critical junctures. Alex Rosenberg and Glenn Cohen read earlier versions of chapters and helped me hone my message in them. Marin Levy and Rahaleh Nassri lent me their ears on more occasions than I should admit and gave me much-needed brain breaks to regain perspective. While many conversations over time have transformed my thinking on the topics in this book, some came at pivotal moments in the development of chapters, including those with Guy Charles, Esther Dyson, Linda Avey, Gavan Fitzsimmons, David Hoffman, Jolynn Dellinger, Emily Murphy, and Laurence Helfer. My colleagues at Duke Law and across the world, and audience members to whom I have presented these ideas asked wise and provocative questions and contributed in many ways big and small.

My research assistants, including Rachel Zacharias and Dylan Stonecipher, enabled me immensely. My assistant, Ava Lane, tirelessly rechecked and formatted the citations, and protected my time to finish writing the book. It takes a village, and our au pairs and nannies over the years have given me the time and pomodoros I needed to focus and write.

Throughout it all, my family—parents, Amir and Afsaneh Farahany, sisters Ava and Amanda, husband Thede, and living children Aristella and Alectra—have been patient, kind, supportive, and given me the much-needed levity, love, and support I needed to persevere. My sister Amanda read many early chapter drafts and provided me with enthusiasm and important feedback. My parents inspired my thinking and developments of many of the stories I used to share my thinking. Our late daughter, Callista, has reshaped my worldview. Her short life and memory have made me a more compassionate human being, and a much better ethicist.

Thanks to all.

Notes

Introduction

1. Michal Teplan, "Fundamentals of EEG Measurement," *Measurement Science Review* 2, no. 2, § 2 (2002): 1–11, http://www.edumed.org.br/cursos/neurociencia/Methods EEGMeasurement.pdf.
2. "All About EMG," Noraxon USA, accessed July 14, 2022, https://www.noraxon.com /all-about-emg/.
3. Yisi Liu, Olga Sourina, and Minh Khoa Nguyen, "Real-Time EEG-Based Emotion Recognition and Its Applications," in *Transactions on Computational Science XII*, ed. Marina L. Gavrilova et al., Lecture Notes in Computer Science, vol. 6670 (Berlin: Springer, 2011): 256–77, https://doi.org/10.1007/978-3-642-22336-5_13.
4. Robin R. Johnson et al., "Drowsiness/Alertness Algorithm Development and Validation Using Synchronized EEG and Cognitive Performance to Individualize a Generalized Model," *Biological Psychology* 87, no. 2 (May 2011): 241–50, https://doi.org /10.1016/j.biopsycho.2011.03.003.
5. Narendra Jadhav, Ramchandra Manthalkar, and Yashwant Joshi, "Effect of Meditation on Emotional Response: An EEG-Based Study," *Biomedical Signal Processing and Control* 34 (2017): 101–13, https://doi.org/10.1016/j.bspc.2017.01.008.
6. Emily A. Vogels, "About One-in-Five Americans Use a Smart Watch or Fitness Tracker," Pew Research Center, January 9, 2020, https://www.pewresearch.org/fact-tank/2020 /01/09/about-one-in-five-americans-use-a-smart-watch-or-fitness-tracker/.
7. "The Rise of mHealth Apps: A Market Snapshot," *Best Practices* (blog), Liquid State, March 26, 2018, updated November 12, 2019, https://liquid-state.com/mhealth-apps -market-snapshot/.
8. Expert Market Research, *Global Neurotechnology Market Report and Forecast 2022–2027*, Report Summary, accessed July 14, 2022, https://www.expertmarketresearch.com /reports/neurotechnology-market.
9. Ned Herrmann, "What Is the Function of the Various Brainwaves?" *Scientific American*,

December 22, 1997, https://www.scientificamerican.com/article/what-is-the-function-of-t-1997-12-22/.

10. Ingrid Johnsen Haas, Melissa N. Baker, and Frank Gonzalez, "Who Can Deviate from the Party Line? Political Ideology Moderates Evaluation of Incongruent Policy Positions in Insula and Anterior Cingulate Cortex," *Social Justice Research* 30, no. 4 (2017): 355–80, https://doi.org/10.1007/s11211-017-0295-0.
11. Julia A. Onton, Dae Y. Kang, and Todd P. Coleman, "Visualization of Whole-Night Sleep EEG from 2-Channel Mobile Recording Device Reveals Distinct Deep Sleep Stages with Differential Electrodermal Activity," *Frontiers in Human Neuroscience* 10 (2016): 605, https://doi.org/10.3389/fnhum.2016.00605.
12. Lisa M. Diamond and Janna A. Dickenson, "The Neuroimaging of Love and Desire: Review and Future Directions," *Clinical Neuropsychiatry* 9, no. 1 (2012), 39–46, https://www.researchgate.net/publication/229424231_The_neuroimaging_of_love_and_desire_Review_and_future_directions.
13. Iris Schutte, J. Leon Kenemans, and Dennis J. L. G. Schutter, "Resting-State Theta/Beta EEG Ratio Is Associated with Reward- and Punishment-Related Reversal Learning," *Cognitive, Affective, & Behavioral Neuroscience* 17, no. 4 (2017): 754–63, https://doi.org/10.3758/s13415-017-0510-3.
14. Andras Horvath et al., "EEG and ERP Biomarkers of Alzheimer's Disease: A Critical Review," *Frontiers in Bioscience-Landmark* 23, no. 2 (January 2018): 183–220, http://dx.doi.org/10.2741/4587.
15. Jennifer J. Newsom and Tara C. Thiagarajan, "EEG Frequency Bands in Psychiatric Disorders: A Review of Resting State Studies," *Frontiers in Human Neuroscience* 12 (2019): 521, https://doi.org/10.3389/fnhum.2018.00521.
16. Knut Engedal et al., "The Power of EEG to Predict Conversion from Mild Cognitive Impairment and Subjective Cognitive Decline to Dementia," *Dementia and Geriatric Cognitive Disorders* 49, no. 1 (2020): 38–47, https://doi.org/10.1159/000508392.
17. "What is the BRAIN Initiative," National Institutes of Health, accessed July 14, 2022, https://braininitiative.nih.gov.
18. Hai Jin, Li-Jun Hou, and Zheng-Guo Wang, "Military Brain Science—How to Influence Future Wars," *Chinese Journal of Traumatology* (in English) 21, no. 5 (2018): 277–80, https://doi.org/10.1016/j.cjtee.2018.01.006.
19. "L'Oréal, in Partnership with Global Neurotech Leader, EMOTIV, Launches New Device to Help Consumers Personalize Their Fragrance Choices," Emotiv, March 21, 2022, https://www.emotiv.com/news/loreal-in-partnership-with-emotiv-neurotech-leader/.
20. *The Autobiography of John Stuart Mill* (Krumlin, UK: Ryburn, 1992).

Chapter 1: The Last Fortress

1. "All About EMG," Noraxon USA, accessed August 17, 2022, https://www.noraxon.com/all-about-emg/.
2. Sara Goering et al., "Recommendations for Responsible Development and Application of Neurotechnologies," *Neuroethics* 14 (2021): 365–86. https://doi.org/10.1007/s12152-021-09468–6.
3. Marcello Ienca and Roberto Adorno, "Towards New Human Rights in the Age of Neuroscience and Neurotechnology," *Life Sciences, Society and Policy,* 13 (2017).
4. Paul Buchheit, "On Gmail, AdSense and More," *BlogScoped,* July 16, 2007, http://blogoscoped.com/archive/2007-07-16-n55.html.
5. *Search Engine Market Share in 2022,* Oberlo, 2022. https://www.oberlo.com/statistics/search-engine-market-share.
6. Ben Gilbert, "How Facebook Makes Money from Your Data, in Mark Zuckerberg's

Words," *Business Insider*, April 11, 2018, https://www.businessinsider.com/how-facebook-makes-money-according-to-mark-zuckerberg-2018-4.

7. Shoshana Zuboff, "Big Other: Surveillance Capitalism and the Prospects of an Information Civilization," *Journal of Information Technology* 30 (2015): 75–89, https://doi.org/10.1057/jit.2015.5.
8. Emil Protalinski, "CTRL-Labs CEO: We'll Have Neural Interfaces in Less Than 5 Years," *VentureBeat*, November 20, 2019, https://venturebeat.com/2019/11/20/ctrl-labs-ceo-well-have-neural-interfaces-in-less-than-5-years/.
9. Elise Reuter, "4 Takeaways from the 23&Me's Planned SPAC Deal," *MedCityNews*, February 7, 2021, https://medcitynews.com/2021/02/four-takeaways-from-23mes-planned-spac-deal/.
10. Charles Seife, "23andMe Is Terrifying, but Not for the Reasons the FDA Thinks," *Scientific American*, November 27, 2013, https://www.scientificamerican.com/article/23andme-is-terrifying-but-not-for-the-reasons-the-fda-thinks/.
11. Emil Protalinski, "Ctrl-Labs CEO: We'll Have Neural Interfaces in Less Than 5 Years," *VentureBeat*, November 20, 2019, https://venturebeat.com/2019/11/20/ctrl-labs-ceo-well-have-neural-interfaces-in-less-than-5-years/.
12. Nick Statt, "Facebook Acquires Neural Interface Startup CTRL-Labs for Its Mind-Reading Wristband," *Verge*, September 23, 2019, https://www.theverge.com/2019/9/23/20881032/facebook-ctrl-labs-acquisition-neural-interface-armband-ar-vr-deal.
13. Rachel Sandler, "Facebook Acquires Brain Computing Startup CTRL Labs," *Forbes*, September 23, 2019, https://www.forbes.com/sites/rachelsandler/2019/09/23/facebook-acquires-brain-computing-startup-ctrl-labs/.
14. Adario Strange, "Facebook Chief Mark Zuckerberg Talks Immersive Remote Video & AR Smartglasses Following Major Reveals from Snap & Google," *Next Reality News*, June 4, 2021, https://next.reality.news/news/facebook-chief-mark-zuckerberg-talks-immersive-remote-video-ar-smartglasses-following-major-reveals-from-snap-google-0384706/.
15. Strange, *Zuckerberg Talks Immersive.*
16. Zongkai Fu, Huiyong Li, Zhenchao Ouyang, Xuefeng Liu, and Jianwei Niu, "Typing Everywhere with an EMG Keyboard: A Novel Myo Armband-Based HCI Tool," *Algorithms and Architectures for Parallel Processing*, Lecture Notes in Computer Science, vol. 12452, ed. Meikang Qiu (Cham, Germany: Springer, 2020), https://doi.org/10.1007/978-3-030-60245-1_17.
17. "Mark Zuckerberg Teases Wearable Tech with Neural Interface in Facebook Post," Reuters, May 4, 2022, https://www.reuters.com/technology/mark-zuckerberg-teases-wearable-tech-with-neural-interface-facebook-post-2022-05-04/.
18. Tommy Palladino, "Facebook's Smartwatch Will Eventually Include CTRL-Labs Tech for Smartglasses Control, Report Says," *Next Reality News*, June 9, 2021, https://next.reality.news/news/facebooks-smartwatch-will-eventually-include-ctrl-labs-tech-for-smartglasses-control-report-says-0384724/.
19. Chris Smith, "Apple Hints AirPods Might Get an Awesome Hidden Feature Soon," *BGR*, June 16, 2021, https://bgr.com/tech/new-airpods-features-sensor-fusion-health-data/.
20. "Enterprise Neurotechnology Solutions," Emotiv, accessed August 18, 2022, https://www.emotiv.com/workplace-wellness-safety-and-productivity-mn8/.
21. Steven Levy, "This Startup Wants to Get in Your Ears and Watch Your Brain," *Wired*, April 14, 2022, https://www.wired.com/story/nextsense-wants-to-get-in-your-ears-and-watch-your-brain/.
22. David Phelan, "AirPods Pro: Apple Hints at Amazing New Direction for Future AirPods," *Forbes*, June 16, 2021, https://www.forbes.com/sites/davidphelan/2021

/06/16/airpods-pro-apple-hints-at-amazing-new-direction-for-future-airpods/?sh=a016cba5cf7a.

23. Brian Heater, "Snap Buys Mind-Controlled Headband Maker, NextMind," *Tech Crunch*, March 23, 2022, https://techcrunch.com/2022/03/23/snap-buys-mind-controlled-headband-maker-nextmind/.
24. Phoebe Weston, "Battle for Control of Your Brain: Microsoft Takes On Facebook with Plans for a Mind-Reading Headband That Will Let You Use Devices with the Power of Thought," *Daily Mail*, January 16, 2018, https://www.dailymail.co.uk/sciencetech/article-5274823/Microsoft-takes-Facebook-mind-reading-technology.html; Cryptocurrency System Using Body Activity Data. US Patent 20200097951, filed September 21, 2018, and issued March 26, 2020, Microsoft Technology Licensing, LLC.
25. Hugo D. Critchley, "Psychophysiology of Neural, Cognitive, and Affective Integration: fMRI and Autonomic Indicants," *International Journal of Psychophysiology* 73, no. 2 (2009): 88–94, https://doi.org/10.1016/j.ijpsycho.2009.01.012.
26. Gillian Grennan et al., "Cognitive and Neural Correlates of Loneliness and Wisdom During Emotional Bias," *Cerebral Cortex* 31, no. 7, (March 2021): 3311–22, https://doi.org/10.1093/cercor/bhab012.
27. Bastian Schiller et al., "Theta Resting EEG in the Right TPJ Is Associated with Individual Differences in Implicit Intergroup Bias," *Social Cognitive and Affective Neuroscience* 14, no. 3, (March 2019): 281–89, https://doi.org/10.1093/scan/nsz007.
28. Mario Frank et al., "Using EEG-Based BCI Devices to Subliminally Probe for Private Information," in *Proceedings of the 2017 on Workshop on Privacy in the Electronic Society* (New York: Association for Computing Machinery, 2017): 133–36, https://doi.org/10.1145/3139550.3139559.
29. Bruno Martín, "Cybersecurity to Guard Against Brain Hacking," *OpenMind BBVA*, January 21, 2020, https://www.bbvaopenmind.com/en/technology/digital-world/cybersecurity-to-guard-against-brain-hacking/.
30. Martin Kaste, "Think Internet Data Mining Goes Too Far? Then You Won't Like This," *All Things Considered*, NPR, May 29, 2014.
31. Ed Brown, "China May Be Using AI to Determine People's Response to 'Thought Education,'" *Newsweek*, July 4, 2022, https://www.newsweek.com/xi-jinping-china-brain-waves-political-thought-education-ccp-ai-1721547.
32. Hayley K. Jach et al., "Decoding Personality Trait Measures from Resting EEG: An Exploratory Report," *Cortex* 130 (September 2020): 158–71, https://doi.org/10.1016/j.cortex.2020.05.013.
33. Sara Goering et al., "Recommendations for Responsible Development and Application of Neurotechnologies," *Neuroethics* 14 (April 2021): 365–86, https://doi.org/10.1007/s12152-021-09468-6.
34. Dimitri Van De Ville et al., "When Makes You Unique: Temporality of the Human Brain Fingerprint," *Science Advances* 7, no. 42 (October 2021), https://doi.org/10.1126/sciadv.abj0751.
35. "IKEA's Bizarre Brain Wave Test for Wannabe Belgian Rug Buyers," *Stuff*, May 24, 2019, https://www.stuff.co.nz/life-style/homed/decor/112983634/ikeas-bizarre-brain-wave-test-for-wannabe-belgian-rug-buyers.
36. *IKEA Heart Scanner*, art event by IKEA, Ogilvy Group, Brussels, Belgium, April 2019, https://www.adforum.com/talent/82210280-gabriel-araujo/work/34594846.
37. Kevin Schoeninger, in discussion with the author, June 24, 2021.
38. "HeartMind Alchemy Lab," Facebook group, https://www.facebook.com/groups/heartmindalchemylab.
39. Mary Madden, "Public Perceptions of Privacy and Security in the Post-Snowden Era," Pew Research Center, Washington, DC (November 12, 2014), https://www.pewresearch.org/internet/2014/11/12/public-privacy-perceptions/.

40. Dulce Ruby, "It's Flowtime: 1st Look at the Biosensing Meditation Tool," *Soul Traveler*, January 7, 2018, https://soultraveler.co/journal/its-flowtime/.
41. Ben Goertzel, "SingularityNET Partners with Chinese Neurotechnology Firm Entertech," *SingularityNet*, November 22, 2018, https://blog.singularitynet.io/singularitynet-partners-with-the-chinese-neurotechnology-firm-entertech-58d25f0a5ecc.
42. "Privacy Policy," Flowtime, last modified June 27, 2019, https://www.meetflowtime.com/privacy-policy.
43. Goertzel, "Chinese Neurotechnology Firm Entertech."
44. Saoirse Kerrigan, "15 Amazing MIT Student Projects of the Past 10 Years," *Interesting Engineering*, June 26, 2018, https://interestingengineering.com/15-amazing-mit-student-projects-of-the-past-10-years.
45. "Multimer Data," Company Profile, Tracxn, last modified June 21, 2022, https://tracxn.com/d/companies/multimerdata.com.
46. "Privacy Policy (Basics)," Mind-Monitor, accessed August 19, 2022, https://mind-monitor.com/FAQ.php#privacy.
47. David Heinzmann, "Home Security Video Captures Close-Up Images of Gunman Killing Companion at Point-Blank Range in Rogers Park," *Chicago Tribune*, July 3, 2020, https://www.chicagotribune.com/fc85e30c-1018-49d7-a2d7-39c422a6d72b-132.html.
48. Nita A. Farahany, "Incriminating Thoughts," *Stanford Law Review* 64, no. 2 (March 2012): 351; Nita A. Farahany, "Searching Secrets," *University of Pennsylvania Law Review* 160 (2012): 1239–1308.
49. Esther Landhuis, "Neuroscience: Big Brain, Big Data," *Nature* 541 (2017): 559–61, https://doi.org/10.1038/541559a.
50. US Army CCDC Army Research Laboratory Public Affairs, "Army Develops Big Data Approach to Neuroscience," US Army, February 5, 2020.
51. Meysam Golmohammadi, Vinit Shah, Iyad Obeid, and Joseph Picone, "Deep Learning Approaches for Automated Seizure Detection from Scalp Electroencephalograms," in *Signal Processing in Medicine and Biology*, ed. Iyad Obeid, Ivan Selesnick, and Joseph Picone (Cham, Switzerland: Springer, 2020), 235–76.
52. US Army CCDC Army Research Laboratory Public Affairs, "Big Data."
53. Louis Henkin, Sarah H. Cleveland, Laurence R. Helfer, Gerald L. Neuman, and Diane F. Orentlicher, *Human Rights: Second Edition* (New York: Foundation Press, 2009).
54. Beth A. Simmons, *Mobilizing for Human Rights: International Law in Domestic Politics* (Cambridge: Cambridge University Press, 2009).
55. Ienca and Adorno, "Towards New Human Rights."
56. Goering et al., "Recommendations for Responsible Development."
57. Birgit Schlütter, *Aspects of Human Rights Interpretation by the UN Treaty Bodies, in UN Human Rights Treaty Bodies: Law and Legitimacy* 261, ed. Helen Keller and Geir Ulfstein (Cambridge, Cambridge University Press, 2012); Rudolf Bernhardt, "Evolutive Treaty Interpretation, Especially of the European Convention on Human Rights," German Y. B. Int'l L., 42:11 (1999).
58. Makue Mutua, *Human Rights Standards: Hegemony, Law and Politics* (Albany: State University of New York Press, 2016).
59. United Nations, Human Rights Council, *Artificial Intelligence and Privacy, and Children's Privacy—Report of the Special Rapporteur on the Right to Privacy*, A/HRC/46/37 (January 25, 2021).
60. Daniel Newman, "Honesty, Transparency and Data Collection: Improving Customer Trust and Loyalty," *Forbes*, February 24, 2021, https://www.forbes.com/sites/danielnewman/2021/02/24/honesty-transparency-and-data-collection-improving-customer-trust-and-loyalty/?sh=206b80d345ab.

Chapter 2: Your Brain at Work

1. "Police Investigate Truck Driver's Facebook Brag About 20 Hours of Driving," *CDL-Life*, November 18, 2015, https://cdllife.com/2015/police-investigate-truck-drivers-facebook-brag-about-20-hours-of-driving/.
2. "Road Transport," *SmartCap*, accessed July 8, 2022, http://www.smartcaptech.com/industries/transport/.
3. Peter Ker, "Australian Employers Are Scanning Their Workers' Minds," *Australian Financial Review*, July 3, 2015, https://www.afr.com/work-and-careers/management/australian-employers-are-scanning-their-workers-minds-20150703-gi46x7; "Wenco International Mining Systems Acquires SmartCap, the World's Leading Fatigue Monitoring Wearable Device," Wenco, May 5, 2021, https://www.wencomine.com/news/wenco-international-mining-systems-acquires-smartcap-the-worlds-leading-fatigue-monitoring-wearable-device.
4. "Wearable Tech Eliminates Microsleeps," *Hazardex*, January 9, 2020, https://www.hazardexonthenet.net/article/176670/Wearable-technology-eliminates-microsleeps.aspx.
5. Cliona McParland and Regina Connolly, "Employee Monitoring in the Digital Era: Managing the Impact of Innovation," *2019 ENTRENOVA Conference Proceedings* (September 2019): 548–57, https://doi.org/10.2139/ssrn.3492245.
6. "Tesco in Trouble over Electronic Armbands," *Triple Pundit*, March 7, 2013, https://www.triplepundit.com/story/2013/tesco-trouble-over-electronic-armbands/107666.
7. Stacy Mitchell, in discussion with the author, December 15, 2020.
8. Ceylan Yeginsu, "If Workers Slack Off, the Wristband Will Know. (And Amazon Has a Patent for It.)," *New York Times*, February 1, 2018, https://www.nytimes.com/2018/02/01/technology/amazon-wristband-tracking-privacy.html.
9. Sarah O'Connor, "Workplace Surveillance May Hurt Us More Than It Helps," *Irish Times*, January 15, 2021, https://www.irishtimes.com/business/work/workplace-surveillance-may-hurt-us-more-than-it-helps-1.4457355.
10. Stacy Mitchell, in discussion with the author, December 15, 2020.
11. "ExpressVPN Survey Reveals the Extent of Surveillance on the Remote Workforce," *ExpressVPN*, May 20, 2021, https://www.expressvpn.com/blog/expressvpn-survey-surveillance-on-the-remote-workforce/.
12. Emine Saner, "Employers Are Monitoring Computers, Toilet Breaks—Even Emotions. Is Your Boss Watching You?," *Guardian* (US edition), May 14, 2018, https://www.theguardian.com/world/2018/may/14/is-your-boss-secretly-or-not-so-secretly-watching-you.
13. "News Roundup, Sept. 24: Fatigue-Detecting Headgear Makes US Debut, Star Chiropractor Cracks Truckers' Backs," *Overdrive Online*, September 24, 2019, https://www.overdriveonline.com/business/article/14896946/news-roundup-sept-24-fatigue-detecting-headgear-makes-us-debut-star-chiropractor-cracks-truckers-backs.
14. SmartCap, *Case Study: Transportation Pilot*, September 17, 2018. http://www.smartcaptech.com/wp-content/uploads/smartcap-case-study-transportation-pilot.pdf.
15. Ker, "Australian Employers Are Scanning"; Chandra Steele, "The Quantified Employee: How Companies Use Tech to Track Workers," *PCMagazine*, February 14, 2020, https://www.pcmag.com/news/the-quantified-employee-how-companies-use-tech-to-track-workers.
16. SmartCap, *Case Study: Hunter Valley Operations (HVO)*, http://www.smartcaptech.com/case-study/hunter-valley-operations-hvo/.
17. BAM Nuttall, "BAM Nuttall and SmartCap Technologies Collaborate to Monitor Construction Workers Fatigue Levels," BAM Nuttall press release, February 16, 2017,

https://www.bam.com/en/press/press-releases/2017/2/bam-nuttall-and-smartcap-technologies-collaborate-to-monitor.

18. SmartCap, *Case Study: Base Metal Mining Operation with Global Interests*, December 13, 2019, http://www.smartcaptech.com/wp-content/uploads/case-study-base-metal-operation-with-global-interest.pdf.
19. SmartCap, *Case Study: Large Copper Mining Operation in East Asia*, October 10, 2019, http://www.smartcaptech.com/wp-content/uploads/case-study-large-copper-mining-operation-in-east-asia.pdf.
20. "Chicago Train Crash Driver Who 'Fell Asleep' Is Sacked," BBC News, April 5, 2014, https://www.bbc.com/news/world-us-canada-26897226.
21. Thomas C. Zambito, "Metro-North Engineer in Fatal Bronx Derailment Drops $10M Lawsuit," *lohud*, August 13, 2019, https://www.lohud.com/story/news/investigations/2019/08/13/metro-north-spuyten-duyvil-lawsuit/2000179001/.
22. Joe Callahan, "Engineer Fell Asleep Before Crash: Recovery Work Remains Ongoing at Crash Site Near Citra," *Ocala StarBanner*, November 22, 2016, https://www.ocala.com/news/20161122/csx-engineer-admits-falling-asleep-before-train-crash.
23. "16 Plane Crashes Caused by Fatigued Aircrew & What It Means for Your Safety-Sensitive Company," https://www.predictivesafety.com/blog/16-plane-crashes-caused-by-fatigued-aircrew.
24. AAA, "Young Drivers Admit to Nodding Off Behind the Wheel," *PR Newswire*, November 9, 2012, https://www.prnewswire.com/news-releases/young-drivers-admit-to-nodding-off-behind-the-wheel-178040441.html.
25. Eric A. Taub, "Sleepy Behind the Wheel? Some Cars Can Tell," *New York Times*, March 16, 2017, https://www.nytimes.com/2017/03/16/automobiles/wheels/drowsy-driving-technology.html.
26. Messaoud Doudou, Abdelmadjid Bouabdallah, and Véronique Berge-Cherfaoui, "Driver Drowsiness Measurement Technologies: Current Research, Market Solutions, and Challenges," *International Journal of Intelligent Transportation Systems Research* 18, no. 2 (2020): Table 1, https://doi.org/10.1007/s13177-019-00199-w.
27. Chris Burt, "New Automotive Biometric and Sensing Technologies Launched by Hyundai, Cadillac, ADI, Yandex, NXP," *BiometricUpdate.com*, March 23, 2020, https://www.biometricupdate.com/202003/new-automotive-biometric-and-sensing-technologies-launched-by-hyundai-cadillac-adi-yandex-nxp.
28. "CoDriver for Driver Monitoring," CoDriver for Car OEMs and Tier-1s, Jungo Connectivity, accessed July 11, 2022, https://www.jungo.com/st/codriver-segments/codriver-driver-monitoring/; Linda Kincaid, "Analog Devices and Jungo Cooperate on In-Cabin Monitoring Technology to Improve Vehicle Safety," Analog Devices, February 5, 2020, https://www.analog.com/en/about-adi/news-room/press-releases/2020/2–5–2020-analog-devices-jungo-cooperate-on-in-cabin-monitoring-technology.html.
29. Paul Sawers, "How Yandex.Taxi is Using Automation to Detect Drowsy and Dangerous Drivers," *VentureBeat*, February 14, 2020, https://venturebeat.com/2020/02/14/how-yandex-taxi-is-using-automation-to-detect-drowsy-and-dangerous-drivers/; Masha Borak, "Didi Detects Drowsy Drivers with AI Facial Recognition," *Abacus*, January 10, 2020, https://www.scmp.com/abacus/news-bites/article/3045621/didi-detects-drowsy-drivers-ai-facial-recognition.
30. Rodney Petrus Balandong et al., "A Review on EEG-Based Automatic Sleepiness Detection Systems for Driver," *IEEE Access* 6 (2018): 22911, https://doi.org/10.1109/ACCESS.2018.2811723.
31. Miankuan Zhu et al., "Vehicle Driver Drowsiness Detection Method Using Wearable EEG Based on Convolution Neural Network," *Neural Computing and Applications* 33,

no. 20 (2021): 13965–80, https://doi.org/10.1007/s00521-021-06038-y; Shubha Majumder et al., "On-Board Drowsiness Detection Using EEG: Current Status and Future Prospects," *2019 IEEE International Conference on Electro Information Technology (EIT)* (May 2019): 483–90, https://doi.org/10.1109/EIT.2019.8833866.

32. John LaRocco, Minh Dong Le, and Dong-Guk Paeng, "A Systemic Review of Available Low-Cost EEG Headsets Used for Drowsiness Detection," *Frontiers in Neuroinformatics* 14 (2020): 42, https://doi.org/10.3389/fninf.2020.553352.
33. Aqsa Mehreen et al., "A Hybrid Scheme for Drowsiness Detection Using Wearable Sensors," *IEEE Sensors Journal* 19, no. 13 (2019): 5119–26, https://doi.org/10.1109/JSEN.2019.2904222; LaRocco, Le, and Paeng, "Low-Cost EEG Headsets"; Miankuan Zhu et al., "Vehicle Driver Drowsiness Detection"; Siwar Chaabene et al., "Convolutional Neural Network for Drowsiness Detection Using EEG Signals," *Sensors* 21, no. 5 (2021): 1734, https://doi.org/10.3390/s21051734; Igor Stancin, Mario Cifrek, and Alan Jovic, "A Review of EEG Signal Features and Their Application in Driver Drowsiness Detection Systems," *Sensors* 21, no. 11 (2021): 3786, https://doi.org/10.3390/s21113786.
34. Khosro Sadeghniiat-Haghighi, Zohreh Yazdi, "Fatigue Management in the Workplace," *Industrial Psychiatry Journal* 24, no. 1 (Jan–Jun2015), https://doi.org/10.4103/0972–6748.160915.
35. Ker, "Australian Employers Are Scanning."
36. Cliona McParland and Regina Connolly, "Employee Monitoring in the Digital Era: Managing the Impact of Innovation," *2019 ENTRENOVA Conference Proceedings* (September 2019): 548–57, https://doi.org/10.2139/ssrn.3492245.
37. Paul J. Zak, "The Neuroscience of Trust: Management Behaviors that Foster Employee Engagement," *Harvard Business Review*, January–February 2017, 84–90.
38. Kristina Martic, "Trust in the Workplace: Why It Is so Important Today and How to Build It," *Haiilo*, accessed July 11, 2022, https://haiilo.com/blog/trust-in-the-workplace-why-it-is-so-important-today-and-how-to-build-it/.
39. Angela Haupt, "Set a Tomato Timer? Eat a Frog? Be Like Ike? Comparing 5 Common Productivity Systems," *Washington Post*, August 2, 2021, https://www.washingtonpost.com/lifestyle/wellness/productivity-pomodoro-gtd-frog-matrix-lee/2021/08/01/0b334dca-efd1–11eb-a452–4da5fe48582d_story.html.
40. Olivier Oullier, "Neuroinformatics: The Real Future of Work" (presentation, Fortune Global Tech Forum, Guangzhou, China, November 7, 2019).
41. Fortune Editors, "These Brain Specialists Have Built Ear Pods to Help You Tune in, Boost Workplace Productivity," *Fortune*, November 7, 2019, https://fortune.com/2019/11/07/brain-ear-pods-boost-productivity-workplace/.
42. Vanessa Micelli-Schmidt and Philip Miseldine, "SAP: Personalizing Workplace Learning with SAP and EMOTIV," *MarketScreener*, September 23, 2019, https://www.marketscreener.com/quote/stock/SAP-SE-436555/news/SAP-nbsp-Personalizing-Workplace-Learning-with-SAP-and-EMOTIV-29239345/.
43. Rebecca J. Compton, Dylan Gearinger, and Hannah Wild, "The Wandering Mind Oscillates: EEG Alpha Power Is Enhanced During Moment of Mind-wandering," *Cognitive, Affective, & Behavioral Neuroscience* 19, no. 5 (2019): 1184–91, https://doi.org/10.3758/s13415-019-00745–9.
44. Deloitte, "Next-Gen Digital Experiences Read Emotions," *Wall Street Journal*, August 6, 2020, https://deloitte.wsj.com/riskandcompliance/2020/08/06/next-gen-digital-experiences-read-emotions/.
45. "CogC2: Cognitive Command and Control," Lockheed Martin, accessed July 11, 2022, https://www.lockheedmartin.com/en-us/capabilities/research-labs/advanced-technology-labs/cog-c2.html.

46. Lily Hay Newman, "The Zoom Privacy Backlash Is Only Getting Started," *Wired,* April 1, 2020, https://www.wired.com/story/zoom-backlash-zero-days/.
47. Mariam Hassib et al., "Brain atWork: Logging Cognitive Engagement and Tasks in the Workplace Using Electroencephalography," in *Proceedings of the 16th International Conference on Mobile and Ubiquitous Multimedia,* November 2017, 305–10, https://doi.org/10.1145/3152832.3152865.
48. Nataliya Kosmyna and Pattie Maes, "AttentivU: An EEG-Based Closed-Loop Biofeedback System for Real-Time Monitoring and Improvement of Engagement for Personalized Learning," *Sensors* 19, no. 23 (2019): 5200
49. Dan Schawbel, "How Covid19 Has Accelerated the Use of Employee Monitoring," *Workplace Intelligence Weekly Newsletter,* August 17, 2020, https://www.linkedin.com/pulse/how-covid-19-has-accelerated-use-employee-monitoring-dan-schawbel/.
50. Alex Tabarrok, "Libertarianism and the Workplace II," *Marginal Revolution,* July 3, 2012.
51. Ker, "Australian Employers Are Scanning."
52. Regulation (EU) 2016/679 of the European Parliament and of the Council of 27 April 2016, on the protection of natural persons with regard to the processing of personal data and on the free movement of such data, and repealing Directive 95/46/EC (General Data Protection Regulation) art. 88, 2016 O.J. L119/1.
53. Ecuador's Constitution, art. 329, https://www.constituteproject.org/constitution/Ecuador_2008.pdf.
54. Chilean Constitution, art. 19, https://www.constituteproject.org/constitution/Chile_2012.pdf.
55. Larry Alton, "Is Constant Corporate Monitoring Killing Morale?" *NBC News,* September 12, 2017, https://www.nbcnews.com/better/business/constant-corporate-monitoring-killing-morale-ncna800301.
56. Cliona McParland and Regina Connolly, "Employee Monitoring in the Digital Era: Managing the Impact of Innovation," *2019 ENTRENOVA Conference Proceedings* (Sept. 2019): 548–557, https://doi.org/10.2139/ssrn.3492245.
57. Jay Greene, "Amazon's Employee Surveillance Fuels Unionization Efforts: 'It's Not Prison, It's Work,'" *Washington Post,* December 2, 2021, https://www.washingtonpost.com/technology/2021/12/02/amazon-workplace-monitoring-unions/.
58. Johana Bhuiyan, "Instacart shoppers say they face unforgiving metrics: 'It's a very easy job to lose,'" *Chicago Tribune,* August 31, 2020, https://www.chicagotribune.com/business/ct-biz-instacart-shoppers-gig-workers-union-20200831-z4hg7ospjjayrfg73a4q7m7224-story.html.
59. Matt Scherer, *Warning: Bossware May Be Hazardous to Your Health* (Washington DC: Center for Democracy and Technology, 2021) https://cdt.org/wp-content/uploads/2021/07/2021-07-29-Warning-Bossware-May-Be-Hazardous-To-Your-Health-Final.pdf.
60. Sara E. Alger, Allison J. Brager, and Vincent F. Capaldi, "Challenging the Stigma of Workplace Napping," *Sleep* 42, no. 8 (Aug 2019): zsz097, https://doi.org/10.1093/sleep/zsz097; Leslie A. Perlow, *Sleeping With Your SmartPhone: How to Break the 24/7 Habit and Change the Way You Work* (Boston: Harvard Business Review Press, 2012).
61. Thorben Lukas Baumgart et al., "Creativity Loading—Please Wait! Investigating the Relationship Between Interruption, Mind Wandering, and Creativity," in *Proceedings of the 53rd Hawaii International Conference on System Sciences* (2020), http://hdl.handle.net/10125/63777.
62. Robert A. Karasek, Jr., "Job Demands, Job Decision Latitude, and Mental Strain: Implications for Job Redesign," *Administrative Science Quarterly* (1979): 285–308.
63. Lydia Smith, "A Week Off to De-Stress: Should Other Companies Follow Bumble's Example?," *Yahoo News,* June 25, 2021, https://news.yahoo.com/as-bumble-gives-staff-a-week-off-to-de-stress-should-other-companies-follow-050027004.html.

64. Lora Jones, "Bumble Closes to Give 'Burnt-Out' Staff a Week's Break," *BBC News*, June 22, 2021, https://www.bbc.com/news/business-57562230.
65. Barnaby Lashbrook, "Is Bumble's Week Off For Burnout The Best Medicine For Stressed Staff?," *Forbes*, June 23, 2021, https://www.forbes.com/sites/barnabylashbrooke/2021/06/23/is-bumbles-week-off-for-burnout-the-best-medicine-for-stressed-staff/?sh=31f649317dcb.
66. Lora Jones, "Bumble Closes to Give 'Burnt-Out' Staff a Week's Break."
67. Joe Pinsker, "Why People Get the 'Sunday Scaries,'" *Atlantic*, February 9, 2020, https://www.theatlantic.com/family/archive/2020/02/sunday-scaries-anxiety-workweek/606289/.
68. Simon Jack, *Lloyds Boss: Mental Health Issues Can Break* Lives, BBC News (Jan. 22, 2020), available at https://www.bbc.com/news/business-51201550.
69. Soeren Mattke, Christopher Schnyer, and Kristin R. Van Busum, *A Review of the US Workplace Wellness Market* (Santa Monica: Rand Corporation, 2012), 5.
70. Rebecca K. Kelly and Sheri Snow, "The Importance of Corporate Wellness Programs for Psychological Health and Productivity in the Workplace," in *Creating Psychologically Healthy Workplaces*, eds. Ronald J. Burke and Atrid M. Richardsen (Cheltenham: Edward Elgar Publishing, 2019), 422.
71. Dylan Haviland, "Aetna's Mindfulness Initiative Leads to Unique Employee Engagement," TTEC, accessed July 11, 2022, https://www.ttec.com/articles/aetnas-mindfulness-initiative-leads-unique-employee-engagement.
72. "Employee Assistance Programs Enhance Corporate Wellness with Brain Training," *SharpBrains*, September 25, 2012, https://sharpbrains.com/blog/2012/09/25/employee-assistance-programs-enhance-corporate-wellness-with-brain-training/#-more-11695.
73. Sarah Wells, "These Robots Want to Read Your Mind While You Work—You Should Let Them," *Inverse*, May 9, 2021, https://www.inverse.com/innovation/mind-reading-robots-are-the-future-of-work.
74. Veronica Sheppard, "Mindfulness Matters: How the Maker of a Brain-Sensing Headband Stays True to Company Values During Difficult Times," *MaRS*, March 5, 2021, https://marsdd.com/news/mindfulness-matters-how-the-maker-of-a-brain-sensing-headband-stays-true-to-company-values-during-difficult-times/.
75. Sheppard, "Mindfulness Matters."
76. "Morneau Shepell and Interaxon Collaborate for Brain Health and Workplace Wellbeing Innovation," *LifeWorks*, January 13, 2020, https://us.lifeworks.com/news/morneau-shepell-and-interaxon-collaborate-brain-health-and-workplace-wellbeing-innovation.
77. Dirk Rodenburg et al., *The Ethics of Brain Wave Technology: Issues, Principles and Guidelines* (CeReB: The Center for Responsible Brainwave Technologies, 2014), https://static1.squarespace.com/static/5344501be4b0d532fc42e22f/t/5390ceece4b0fe2199de93cc/1401999084766/The+Ethics+of+Brainwave+Technolo%20gy.pdf.
78. "Legal," Muse, last modified June 24, 2022, https://choosemuse.com/legal/.
79. "Legal," Muse.
80. "Wellbeing Services," Thought Beanie, accessed July 11, 2022, http://www.thoughtbeanie.com/corporate.
81. "Enterprise Neurotechnology Solutions," accessed July 14, 2022, https://www.emotiv.com/workplace-wellness-safety-and-productivity-mn8/.
82. Timothy Gubler, Ian Larkin, and Lamar Pierce, "Doing Well by Making Well: The Impact of Corporate Wellness Program on Employee Productivity," *Management Science* 64, no. 11 (June 2018), 4967–4987, https://doi.org/10.1287/mnsc.2017.2883. Healthier employees are not only less expensive, but also significantly more productive. One study compared the productivity of employees at an industrial laundry

company that offered an employee wellness program at 4 out of 5 of its locations. The sites that offered the wellness program enjoyed a 4 percent increase in worker productivity, or one additional productive workday per month per employee.

83. "HIPAA Privacy and Security and Workplace Wellness Programs," HHS.gov, US Department of Health & Human Services, last reviewed April 20, 2015, https://www.hhs.gov/hipaa/for-professionals/privacy/workplace-wellness/index.html.
84. Ifeoma Ajunwa, Kate Crawford, and Jason Schultz, "Limitless Workers Surveillance," *California Law Review* 105(3): 735–776 (June 2017).
85. Christopher Rowland, "With Fitness Trackers in the Workplace, Bosses Can Monitor Your Every Step—And Possibly More," *Washington Post*, February 16, 2019.
86. Rowland, "With Fitness Trackers in the Workplace, Bosses Can Monitor Your Every Step."
87. Rachel Emma Silverman, "Bosses Tap Outside Firms to Predict Which Workers Might Get Sick," *Wall Street Journal*, February 17, 2016. For example, Wal-Mart Stores, Inc., contracts with Castlight Healthcare Inc. to crunch employee data from Wal-Mart's wellness program.
88. Hannah-Kaye Fleming, "Navigating Workplace Wellness Programs in the Age of Technology and Big Data," *Journal of Science Policy & Governance* 17, no. 1 (Sept 2020), https://doi.org/10.38126/JSPG170104.
89. Justin Sherman, *Data Brokers and Sensitive Data on U.S. Individuals: Threats to American Civil Rights, National Security, and Democracy*, Duke Sanford Cyber Policy Program, August 2021, https://sites.sanford.duke.edu/techpolicy/wp-content/uploads/sites/17/2021/08/Data-Brokers-and-Sensitive-Data-on-US-Individuals-Sherman-2021.pdf.
90. Houtan Jebelli, Sungjoo Hwang, and SangHyun Lee, "EEG based Workers' Stress Recognition at Construction Sites," *Automation in Construction* 93 (2018): 315–324, https://doi.org/10.1016/j.autcon.2018.05.027.
91. Alan Ferguson, "Ready to Wear: Wearable Technology Could Boost Workplace Safety, But Concerns Remain," *Saftey+Health*, February 24, 2019, https://www.safetyandhealthmagazine.com/articles/18093-ready-to-wear-wearable-technology-could-boost-workplace-safety-but-concerns-remain.
92. F. M. Al-Shargie et al., "Mental stress quantification using EEG signals," *International Conference for Innovation in Biomedical Engineering and Life Sciences*, ed. Fatima Ibrahim et al.(Singapore: Springer, 2015), 15–19. https://doi.org/10.1007/978–981–10–0266–3_4.
93. Dan Schawbel, "How Covid-19 has Accelerated the Use of Employee Monitoring," *Workplace Intelligence Weekly Newsletter*, August 17, 2020, https://www.linkedin.com/pulse/how-covid-19-has-accelerated-use-employee-monitoring-dan-schawbel/.
94. Emine Saner, "Employers are monitoring computers, toilet breaks—even emotions. Is your boss watching you?," *Guardian*, May 14, 2018, https://www.theguardian.com/world/2018/may/14/is-your-boss-secretly-or-not-so-secretly-watching-you. One study monitored the emails, management interactions, and productivity of men and women at work, found that men and women had nearly identical objective behaviors in the workplace. Which challenged a longstanding view that women received fewer promotions and lower wages due to their behavior or lesser productivity. Greater workplace equality could follow from what we learn.
95. "Workplace Privacy," ACLU, accessed July 11, 2022, https://www.aclu.org/issues/privacy-technology/workplace-privacy.
96. Cliona McParland and Regina Connolly, "Employee Monitoring in the Digital Era: Managing the Impact of Innovation," *2019 ENTRENOVA Conference Proceedings* (Sept. 2019): 548–557, https://doi.org/10.2139/ssrn.3492245.
97. Research studies have shown EEG can be highly accurate for emotion detection, and that in-ear EEG can achieve high levels of accuracy, as well. See., e.g. Chanavit

Athavipach, Setha Pan-Ngum, and Pasin Israsena, "A Wearable In-Ear EEG Device for Emotion Monitoring," *Sensors* 19, no. 18 (2019): 4014; Gang Li, Zhe Zhang, and Guoxing Wang, "Emotion Recognition Based on Low-Cost In-Ear EEG," in *2017 IEEE Biomedical Circuits and Systems Conference (BioCAS),* (2017): 1–4.

98. Benjamin H. Harris, "Fostering More Competitive Labor Markets through Transparent Wages," *Inequality and the Labor Market: The Case for Greater Competition,* ed. Sharon Block and Benjamin H. Harris (Washington, DC: Brookings Institution Press, 2021), 149–158.
99. Thomas R. Sadler and Shane Sanders, "The 2011–2021 NBA Collective Bargaining Agreement: Asymmetric Information, Bargaining Power and the Principal Agency Problem," *Managerial Finance* 42, no. 9 (2016): 891–901, https://doi.org/10.1108/MF-02-2016-0048.
100. David Aboody and Baruch Lev, "Information Asymmetry, R&D, and Insider Gains," *The Journal of Finance* 55, no. 6 (2000): 2747–66.
101. Amir Sufi, "Information Asymmetry and Financing Arrangements: Evidence from Syndicated Loans," *Journal of Finance* 62, no. 2 (2007): 629–68, https://doi.org/10.1111/j.1540-6261.2007.01219.x.
102. Jerrin Thomas Panachakel and Angarai Ganesan Ramakrishnan, "*Decoding Covert Speech from EEG-A Comprehensive Review,*" *Frontiers in Neuroscience* 15, no. 642251 (Apr 2021), https://doi.org/10.3389/fnins.2021.642251.
103. Mike Butcher, "Employee Talent Predictor Retrain.Ai Raised Another $7M, Adds Splunk as Strategic Investor," *Yahoo News,* August 13, 2021, https://news.yahoo.com/employee-talent-predictor-retrain-ai-155538602.html.
104. Rohini Das et al., "Brain Signal Analysis for Mind Controlled Type-Writer Using a Deep Neural Network," in *2020 International Conference on Wireless Communications Signal Processing and Networking (WiSPNET),* (IEEE, 2020): 149–153, https://doi.org/10.1109/WiSPNET48689.2020.9198330; Sayantani Ghosh et al., "Vowel Sound Imagery Decoding by a Capsule Network for the Design of an Automatic Mind-Driven Type-Writer," in *2020 International Joint Conference on Neural Networks (IJCNN),* (IEEE, 2020): 1–8, https://doi.org/10.1109/IJCNN48605.2020.9206754.
105. Efy Yosrita et al., "EEG Based Identification of Words on Exam Models with Yes-No Answers for Students with Visual Impairments," in *2019 IEEE International Conference on Engineering, Technology and Education (TALE),* (IEEE, 2019): 1–5, https://doi.org/10.1109/TALE48000.2019.9225903.
106. Suzanne Dikker et al., "Brain-to-Brain Synchrony Tracks Real-World Dynamic Group Interactions in the Classroom," *Current Biology* 27, no. 9 (May 2017): 1375–1380, https://doi.org/10.1016/j.cub.2017.04.002.
107. Dikker et al., "Brain-to-Brain Synchrony."
108. Stacy Mitchell, in discussion with the author, December 15, 2020.
109. "Employees Accuse Google of Surveillance at Workplace," *Hindu,* October 25, 2019, https://www.thehindu.com/sci-tech/technology/employees-accuse-google-of-surveillance-at-workplace/article29795890.ece.
110. Charlotte Garden, "Labor Organizing in the Age of Surveillance," *Saint Louis University Law Journal* 63 (2018): 59, citing Colgate-Palmolive Co., 323 N.L.R.B. 515, 517 (1997).
111. Lisa Marie Segarra, "More Than 20,000 Google Employees Participated in Walkout Over Sexual Harassment Policy," *Fortune,* November 3, 2018, https://fortune.com/2018/11/03/google-employees-walkout-demands/.; Chris Ayres, "The Truth About Andy Rubin and Google's Existential Crisis," *GQ,* March 12, 2020, https://www.gq-magazine.co.uk/politics/article/andy-rubin-google.
112. "Workplace Privacy," ACLU, accessed July 11, 2022, https://www.aclu.org/issues/privacy-technology/workplace-privacy.

113. Ker, "Australian Employers Are Scanning."
114. Sophie Chapman, "BHP Billiton Using Caps to Monitor Drivers' Brainwaves," *Mining Digital*, May 17, 2020, https://miningdigital.com/technology/bhp-billiton-using-caps-monitor-drivers-brainwaves.
115. Kateryna Maltseva, "Wearables in the Workplace: The Brave New World of Employee Engagement," *Business Horizons* 63, no. 4 (July-Aug 2020): 493–505, https://doi.org/10.1016/j.bushor.2020.03.007.

Chapter 3: Big Brother Is Listening

1. Ellen Ioanes, "What Peng Shuai's Rape Accusation Says About China," *Vox*, December 5, 2021, https://www.vox.com/2021/12/5/22818595/peng-shuai-disappearance-china-me-too-sexual-assault.
2. Paul Mozur et al., "Beijing Silenced Peng Shuai in 20 Minutes, Then Spent Weeks on Damage Control," *New York Times*, December 8, 2021, https://www.nytimes.com/interactive/2021/12/08/world/asia/peng-shuai-china-censorship.html.
3. Women's Tennis Assosciation, "Steve Simon Announces WTA's Decision to Suspend Tournaments in China," press statement, December 3, 2021, https://www.wtatennis.com/news/2384758/steve-simon-announces-wta-s-decision-to-suspend-tournaments-in-china.
4. Andrés Martinez, "Future Tense Newsletter: Tennis Diplomacy and Internet Freedom," *Slate*, December 4, 2021, https://slate.com/technology/2021/12/future-tense-newsletter-peng-shuai-internet-freedom.html.
5. Ben Church, "Peng Shuai: Human Rights Activist Peter Dahlin Says IOC Is Putting Chinese Tennis Star at 'Greater Risk,'" CNN, December 3, 2021, https://www.cnn.com/2021/12/03/sport/ioc-peter-dahlin-peng-shuai-spt-intl/index.html.
6. Riley Morgan, "Peng Shuai Detail Sparks 'Unacceptable' Controversy at Wimbledon," *Yahoo Sport Australia*, July 4, 2022, https://au.sports.yahoo.com/wimbledon-2022-telling-peng-shuai-detail-sparks-outrageous-furore-020459825.html.
7. "IOC Says Olympic Athletes Will Be Safe and Free to Express Opinions, Within the Law, at Beijing Olympics," *MassNews*, December 10, 2021, https://www.massnews.com/ioc-says-olympic-athletes-will-be-safe-and-free-to-express-opinions-within-the-law-at-beijing-olympics/.
8. Sara Fischer, "Russia's Crackdown on Free Press and Speech Intensifies," *Axios*, March 5, 2022, https://www.axios.com/russia-crackdown-press-social-media-information-war-287c0e56-be3e-4bd9-af47–037603ee859d.html; Shirin Ghaffary, "Russia Continues Its Online Censorship Spree by Blocking Instagram," *Vox*, March 11, 2022, https://www.vox.com/recode/22962274/russia-block-facebook-restrict-twitter-putin-censorship-ukraine.
9. John Stuart Mill, *On Liberty* (London: John W. Parker and Son, 1859): 22.
10. Steve Cannane, "Vladimir Putin's Next War Is Being Waged Against Russia's Independent Media," ABC News Australia, March 6, 2022, https://www.abc.net.au/news/2022–03–07/vladimir-putin-cracks-down-on-free-speech-with-new-laws/100887310.
11. Stephen Chen, "'Forget the Facebook Leak': China Is Mining Data Directly from Workers' Brains on an Industrial Scale," *South China Morning Post*, April 29, 2018, https://www.scmp.com/news/china/society/article/2143899/forget-facebook-leak-china-mining-data-directly-workers-brains.
12. Stephen Chen, "'Forget the Facebook Leak': China Is Mining Data Directly from Workers' Brains on an Industrial Scale," *South China Morning Post*, April 29, 2018, https://www.scmp.com/news/china/society/article/2143899/forget-facebook-leak-china-mining-data-directly-workers-brains.
13. Yifan Wang, Shen Hong, and Crystal Tai, "China's Efforts to Lead the Way in AI Start

in Its Classroms," *Wall Street Journal*, October 24, 2019, https://www.wsj.com/articles/chinas-efforts-to-lead-the-way-in-ai-start-in-its-classrooms-11571958181.
14. Jane Li, "A 'Brain- Reading' Headband for Students Is Too Much Even for Chinese Parents," *Quartz*, November 5, 2019, https://qz.com/1742279/a-mind-reading-headband-is-facing-backlash-in-china/.
15. Wang, Hong, and Tai, "China's Efforts to Lead."
16. Kezia Parkins, "Primary School in China Suspends Use of BrainCo Brainwave Tracking Headband," Global Shakers, November 6, 2019, https://globalshakers.com/primary-school-in-china-suspends-use-of-brainco-brainwave-tracking-headband/.
17. "Support INCF's Mission to Push Global Neuroscience Further and Faster," International Neuroinformatics Coordinating Facility, accessed July 12, 2022, https://www.incf.org/donate.
18. Summer Allen, "What Exactly Is Obama's $100 Million BRAIN Initiative," AAAS, April 15, 2013, https://www.aaas.org/what-exactly-obamas-100-million-brain-initiative.
19. Andy Coravos, "The BRAIN Initiative: The Race to Understand the Human Brain," *NeuroTechX Content Lab* (blog), Medium, April 12, 2017, https://medium.com/neurotechx/the-brain-initiative-the-race-to-understand-the-human-brain-c2fddfea271e.
20. Robin Marks, "'Neuroprosthesis' Restores Words to Man with Paralysis," UCSF News & Media, University of California, San Francisco, July 14, 2021, https://www.ucsf.edu/news/2021/07/420946/neuroprosthesis-restores-words-man-paralysis.
21. Sten Grillner et al., "Worldwide Initiatives to Advance Brain Research," *Nature Neuroscience* 19, no. 9 (2016): 1118–22, https://doi.org/10.1038%2Fnn.4371.
22. International Brain Initiative, "International Brain Initiative: An Innovative Framework for Coordinated Global Brain Research Efforts," *Neuron* 105, no. 2 (January 2020): 212–16, https://doi.org/10.1016/j.neuron.2020.01.002.; International Brain Initiative, "Brain Initiatives Move Forward Together," press release, December 11, 2017, https://www.internationalbraininitiative.org/brain-initiatives-move-forward-together; Shelly Fan, "Decoding the Brain Goes Global with the International Brain Initiative," *SingularityHub*, January 28, 2020, https://singularityhub.com/2020/01/28/decoding-the-brain-goes-global-with-the-international-brain-initiative/.
23. Barack Obama to Amy Gutmann, Letter to the Chair of the Presidential Commission on the Study of Bioethical Issues, July 1, 2013, in *Gray Matters: Topics at the Intersection of Neuroscience, Ethics, and Society* 2 (March 2015): vi–vii.
24. United States, Presidential Commission for the Study of Bioethical Issues, *Gray Matters: Topics at the Intersection of Neuroscience, Ethics, and Society*, vol. 2, (Washington, DC: Presidential Commission for the Study of Bioethical Issues, 2015).
25. Fan, "Decoding the Brain."
26. Adam Kolber, "Freedom of Memory Today," *Neuroethics* 2 (2008):145–48.
27. Council of Europe, *European Convention on Human Rights* (Strasbourg: Council of Europe, 2021), https://www.echr.coe.int/documents/convention_eng.pdf.
28. United Nations, General Assembly, *Interim Report of the Special Rapporteur on Freedom of Religion or Belief, Ahmed Shaheed*, A/75/380 (October 5, 2021), https://documents-dds-ny.un.org/doc/UNDOC/GEN/N21/274/90/PDF/N2127490.pdf.
29. Simon McCarthy-Jones, "The Autonomous Mind: The Right to Freedom of Thought in the Twenty-First Century," *Frontiers in Artificial Intelligence* 2 (2019): 19, https://doi.org/10.3389/frai.2019.00019; J. Cardozo, *Palko v. Connecticut*, 302 U.S. 319 (1937) (Freedom of thought . . . is the matrix, the indispensable condition, of nearly every other freedom. With rare aberrations a pervasive recognition of this truth can be traced in our history, political and legal.); see, for example, *Stanley v. Georgia*, 394 U.S. 557 (1969).
30. *West Virginia State Board of Education v. Barnett*, 319 U.S. 624 (1943).
31. The most common context in which freedom of thought has arisen in US law is

when a criminal defendant challenges their conviction for an attempted but not completed crime. Take the sordid case of *US v. Ganaches*, 156 F.3d 1 (1st Cir. 1998), which involved the attempted sexual assault of a minor in 1995. In that case, a police detective in Keene, New Hampshire, set up a sting operation to try to find and arrest would-be child molesters. Posing as the mother of young children who was seeking a male partner to engage in sex with them while she photographed them, the detective put an advertisement in a "swingers" magazine. The defendant responded to the advertisement and engaged in several months of correspondence. Finally, he traveled across state lines to New Hampshire, and waited in a motel parking lot for the mother and her three children to arrive. Instead, he was met by the detective and was arrested and ultimately convicted for the attempted sexual assault of a minor. In his appeal, the defendant argued that he was being punished for "mere thought" and not conduct, since he was arrested before he could carry out his intentions. The court considered his claim and agreed that freedom of thought protects individuals from being arrested and convicted of a crime for merely *thinking* about committing an illegal act. But he lost his appeal because he didn't just "sit in the quiet of his house" while contemplating "evil thought." Instead, he took substantial steps toward committing the crime and that was what he was being punished for.

32. Nita A. Farahany, "The Costs of Changing our Minds," *Emory Law Journal* 69 (2019): 75–110.
33. Jan Christoph Bublitz, "Freedom of Thought in the Age of Neuroscience: A Plea and a Proposal for the Renaissance of a Forgotten Fundamental Right," *Archives for Philosophy of Law and Social Philosophy* 100, no. 1 (2014): 1–25, https://www.jstor.org/stable/24756752.
34. Bublitz, "Age of Neuroscience."
35. B. P. Vermeulen, "Freedom of Thought, Conscience and Religion (Article 9)," in *Theory and Practice of the European Convention on Human Rights*, ed. Pieter van Dijk et al., 4th ed., (Cambridge: Intersentia Publishers, 2006): 751–71. The framework builds on that builds upon the one proposed by Professor B. P. Vermeulen in 2006, in his nonbinding commentary on Article 9 of the EHCR.
36. Vermeulen, "Freedom of Thought, Conscience"; "Advance Unedited Version: The Freedom of Thought—Note by the Secretary-General," submitted to the Seventy-sixth session of the UN General Assembly, A/75/380 (October 5, 2021).
37. Ronen Kopito et al., "Brain-Based Authentication: Towards a Scalable, Commercial Grade Solution Using Noninvasive Brain Signals," *bioRxiv* (April 30, 2021), https://doi.org/10.1101/2021.04.09.439244; Sooriyaaarachchi et al., "MusicID."
38. Dimitri Van de Ville et al., "When Makes You Unique: Temporality of the Human Brain Fingerprint," *Science Advances* 7, no. 42 (2021): eabj0751, https://doi.org/10.1126/sciadv.abj0751.
39. Amir Jalaly Bigdoly, Hamed Jalaly Bidgoly, and Zeynab Arezoumand, "A Survey on Methods and Challenges in EEG-Based Authentication," *Computers & Security* 93 (June 2020): 101788.
40. Jinani Sooriyaaarachchi et al., "MusicID: A Brainwave-Based User Authentication System for Internet of Things," *IEEE Internet of Things Journal* 8, no. 10 (May 15, 2021): 8304–13, https://doi.org/ 10.1109/JIOT.2020.3044726.
41. "Biometrics in Government: Enhanced Security and Convenience for Citizens," Aware, Inc., accessed July 12, 2022, https://www.aware.com/blog-biometrics-in-government-enhanced-security/.
42. "Mandatory National IDs and Biometric Databases," Electronic Frontier Foundation, accessed July 12, 2022, https://www.eff.org/issues/national-ids.
43. Office of the Privacy Commissioner of Canada, "Data at Your Fingertips Biometrics and the Challenges to Privacy," published February 2011, updated March 2, 2022, https:

//www.priv.gc.ca/en/privacy-topics/health-genetic-and-other-body-information/gd_bio_201102/.

44. Elizabeth Neus, "How the Government Uses Biometric Authentication Technology," *FedTech Magazine*, August 30, 2019, https://fedtechmagazine.com/article/2019/08/how-government-uses-biometric-authentication-technology-perfcon.
45. Kopito et al., "Brain-Based Authentication."
46. Paul Bischoff, "Biometric Data: 100 Countries Ranked by How They're Collecting It and What They're Doing With It," Comparitech, last modified April 4, 2022, https://www.comparitech.com/blog/vpn-privacy/biometric-data-study/.
47. Krisztina Huszti-Orbán and Fionnuala Ní Aoláin, "Use of Biometric Data to Identify Terrorists: Best Practice or Risky Business?," University of Minnesota Human Rights Center (Minneapolis: Regents of the University of Minnesota, 2020), https://www.ohchr.org/Documents/Issues/Terrorism/Use-Biometric-Data-Report.pdf.
48. Aaron Boyd, "White House Wants to Know How Biometrics Like Facial Recognition Are Being Used," *Nextgov*, October 12, 2021, https://www.nextgov.com/emerging-tech/2021/10/white-house-wants-know-how-biometrics-facial-recognition-are-being-used/186033/.
49. Marcy Mason, "Biometric Breakthrough: How CBP Is Meeting Its Mandate and Keeping America Safe," US Customs and Border Patrol, last modified January 4, 2022, https://www.cbp.gov/frontline/cbp-biometric-testing.
50. Daine Taylor, "How Facial Recognition Technology Is Being Used at Airports," *Travel Market Report*, July 18, 2019, https://www.travelmarketreport.com/articles/How-Facial-Recognition-Technology-is-Being-Used-at-Airports.; US Department of Homeland Security, "Proposed Rules: Collection of Biometric Data from Aliens upon Entry to and Departure from the United States," *Federal Register* 85, no. 224 (November 19, 2020): 74162, https://www.govinfo.gov/content/pkg/FR-2020–11–19/pdf/2020–24707.pdf.
51. See, for example, Haonan Ren et al., "Identifying Individuals by fNIRS-Based Brain Functional Network Fingerprints," *Frontiers in Neuroscience* 16 (February 2022), https://doi.org/10.3389/fnins.2022.813293.
52. Sooriyaarachchi et al., "MusicID."
53. Sooriyaarachchi et al.
54. Seyed Abolfazle Valizadeh et al., "Decrypting the Electrophysiological Individuality of the Human Brain: Identification of Individuals Based on Resting-State EEG Activity," *NeuroImage* 197 (2019): 470–81, https://doi.org/10.1016/j.neuroimage.2019.04.005.
55. Joseph Lange et al. "Side-Channel Attacks Against the Human Brain: The PIN Code Case Study (extended version)," *Brain Informatics* 5, no. 12 (October 2018): 1–16, https://doi.org/10.1186/s40708-018-0090–1.
56. Michael Inzlicht et al., "Neural Markers of Religious Conviction," *Psychological Science* 20, no. 3 (2009): 385–92, https://doi.org/10.1111/j.1467–9280.2009.02305.x.
57. Elizabeth Stoycheff, "Under Surveillance: Examining Facebook's Spiral of Silence Effects in the Wake of NSA Internet Monitoring." *Journalism & Mass Communication Quarterly* 93, no. 2 (2016): 296–311, https://doi.org/10.1177%2F1077699016630255.
58. Elisabeth Noelle-Neumann, *The Spiral of Silence: Public Opinion—Our Social Skin*, 2nd ed. (Chicago: University of Chicago Press, 1993).
59. Karen Turner, "Mass Surveillance Silences Minority Opinions, According to Study," *Washington Post*, March 28, 2016, https://www.washingtonpost.com/news/the-switch/wp/2016/03/28/mass-surveillance-silences-minority-opinions-according-to-study/.
60. Simon McCarthy-Jones, "The Autonomous Mind: The Right to Freedom of Thought in the Twenty-First Century," *Frontiers in Artificial Intelligence* 2 (2019): 19, https://doi.org/10.3389/frai.2019.00019.

61. Stanley Milgram, *Obedience to Authority: An Experimental View* (New York: Harper & Row, 1974).
62. Michael Kenneth Isenman, review of *Crimes of Obedience: Toward Social Psychology of Authority and Responsibility*, by Herbert C. Kelman and V. Lee Hamilton, *Michigan Law Review* 88, no. 6 (1990): 1474.
63. S. Alexander Haslam and Stephen D. Reicher, "Contesting the 'Nature' of Conformity: What Milgram and Zimbardo's Studies Really Show," *PLOS Biology* 10, no. 11 (2012): e1001426, https://doi.org/10.1371/journal.pbio.1001426.
64. Mill, *On Liberty*, 34.
65. Simon McCarthy-Jones, "Freedom of Thought Is Under Attack—Here's How to Save Your Mind," *Conversation*, October 21, 2019, https://theconversation.com/freedom-of-thought-is-under-attack-heres-how-to-save-your-mind-124379.
66. David Kocieniewski and Peter Robison, "Trump Aide Partnered with Firm Run by Man with Alleged KGB Ties," *Bloomberg*, December 23, 2016, https://www.bloomberg.com/news/articles/2016-12-23/trump-aide-partnered-with-firm-run-by-man-with-alleged-kgb-ties.
67. Jerry Markon, "Michael Flynn Had Role in Firm Co-led by Man Who Tried to Sell Material to the KGB," *Chicago Tribune*, December 23, 2016, https://www.chicagotribune.com/nation-world/ct-trump-flynn-brainwave-science-kgb-20161223-story.html.
68. Markon, "Michael Flynn Had Role in Firm."
69. Arab News, "What Is Brain Fingerprinting?" *Labworx*, October 5, 2021, https://www.arablab.com/file_store/news-item/364/Arab%20News%20-%20What%20is%20Brain%20Fingerprinting%20(V2%20-%2010.05.21).pdf.
70. Ayub Dawood, "Dubai Police Use Brain Waves to Look Inside Suspect's Mind to Solve Murder Case," *Mashable*, January 26, 2021, https://me.mashable.com/tech/12683/dubai-police-use-brain-waves-to-look-inside-suspects-mind-to-solve-murder-case.
71. Samuel Sutton et al., "Evoked-Potential Correlates of Stimulus Uncertainty," *Science* 150, no. 3700 (1965): 1187–88, https://doi.org/10.1126/science.150.3700.1187.
72. Tim Stelloh, "Larry Farwell Claims His Lie Detector System Can Read Your Mind. Is He a Scam Artist, or a Genius?," *OneZero* (blog), Medium, January 6, 2021, https://onezero.medium.com/larry-farwell-claims-his-lie-detector-system-can-read-your-mind-is-he-a-scam-artist-or-a-genius-2aebd4e041ca.
73. Lawrence A. Farwell and Emanuel Donchin, "The Truth Will Out: Interrogative Polygraphy ('Lie Detection') with Event-Related Brain Potentials," *Psychophysiology* 28, no. 5 (1991): 531–47.
74. Nita A. Farahany, "Incriminating Thoughts," *Stanford Law Review* 64 (2012): 351–408.
75. Sarah Sturman Dale, "The Brain Scientist: Climbing Inside the Criminal Mind," *Time*, November 26, 2001, https://content.time.com/time/subscriber/article/0,33009,1001318,00.html.
76. Stelloh, "Larry Farwell Claims."
77. *See, Johnson v. State*, 730 N.W.2d 209 (Ct. App. Iowa 2007); *People v. Dorris*, 2013 IL App (4th) 120699-U (App. Ct. Ill. 2013); E.g., *State v. Bates*, 2007 WL 2580552 (Superior Ct. N.J. 2007).
78. *State v. Harrington*, 284 N.W.2d 244 (Iowa 1979).
79. Lawrence A. Farwell and Thomas H. Makeig, "Farwell Brain Fingerprinting in the Case of Harrington v. State," *Open Court* 10, no. 3 (2005): 7–10, https://larryfarwell.com/pdf/OpenCourtFarwellMakeig-dr-larry-farwell-brain-fingerprinting-dr-lawrence-farwell.pdf.
80. Jayanth Murali, "Cool Tool for Police Investigation?," *Deccan Chronicle*, September 3, 2018, https://www.deccanchronicle.com/nation/current-affairs/030918/cool-tool-for-police-investigation.html.
81. Letter from Robert D. Dawson, Macon County Sheriff, Macon, MO, March 2,

2002, https://larryfarwell.com/pdf/Grinder-Dawson-Letter-dr-larry-farwell-brain-fingerprinting-dr-lawrence-farwell.pdf.

82. David Cox, "Can Your Brain Reveal You Are a Liar?," BBC, January 25, 2016, https://www.bbc.com/future/article/20160125-is-it-wise-that-the-police-have-started-scanning-brains.
83. K. V. Dijkstra, J. D. R. Farquhar, and P. W. M. Desain, "The N400 for Brain Computer Interfacing: Complexities and Opportunities," *Journal of Neural Engineering* 17, no. 2 (2020): 022001, https://doi.org/10.1088/1741–2552/ab702e.
84. Joseph R. Simpson, "Functional MRI Lie Detection: Too Good to Be True?," *Journal of the American Academy of Psychiatry and the Law* 36, no. 4 (2008): 491, 492–93. These include the anterior cingulate cortex, which is involved in attention and monitoring processes, and the left dorsolateral and right anterior prefrontal cortices, areas of executive function involved in working memory and behavioral control. It has been hypothesized that these regions are recruited for the purpose of inhibiting a prepotent response (the truth) while simultaneously constructing new information (the lie).
85. Associated Press, "Murder Trial with Fitbit Evidence Heads to Jury Selection," *U.S. News*, February 28, 2022, https://www.usnews.com/news/best-states/connecticut/articles/2022–02–28/murder-trial-with-fitbit-evidence-heads-to-jury-selection.
86. Erin Moriarty, "21st Century Technology Used to Help Solve Wisconsin Mom's Murder," CBS News, October 20, 2018, https://www.cbsnews.com/news/the-fitbit-alibi-21st-century-technology-used-to-help-solve-wisconsin-moms-murder/.
87. Leah Burrows, "To Be Let Alone: Brandeis Foresaw Privacy Problems," *BrandeisNOW*, Brandeis University, July 24, 2013, https://www.brandeis.edu/now/2013/july/privacy.html.
88. Samuel Warren and Louis Brandeis, "The Right to Privacy," *Harvard Law Review* 4, no. 5 (1890): 193–220.
89. *Stanley v. Georgia*, 394 U.S. 557 (1969).
90. *Olmstead v. United States*, 277 U.S. 438, 474 (1928).
91. Mill, *On Liberty*, 33–34.
92. UN Human Rights Committee, *CCPR General Comment No. 22: Article 18 (Freedom of Thought, Conscience or Religion)*, CCPR/C/21/Rev.1/Add.4 (July 30, 1993), https://www.refworld.org/docid/453883fb22.html.

Chapter 4: Know Thyself

1. Alyssa Barbieri, "Some Scouts Had Justin Fields Rated as QB1 in this Draft Class," *USA Today*, May 3, 2021, https://bearswire.usatoday.com/2021/05/03/bears-justin-fields-top-quarterback-draft-class/.
2. Joey Kaufman, "Ohio State Quarterback Justin Fields Speaks about Epilepsy Diagnosis for the First Time," *Buckeye Xtra*, April 30, 2021, https://www.buckeyextra.com/story/football/2021/04/30/ohio-state-quarterback-justin-fields-discusses-epilepsy-diagnosis/4893387001/.
3. Steve Johnson, "Epilepsy community says Chicago Bears' Justin Fields revealing his condition is landmark moment," *Chicago Tribune*, May 14, 2021, https://www.chicagotribune.com/living/health/ct-bears-justin-fields-epilepsy-community-health-reaction-20210514-w5k5ozs6zzbufnyx5rpen6xbg4-story.html.
4. World Health Organization, *Epilepsy Fact Sheet*, last modified February 9, 2022, https://www.who.int/news-room/fact-sheets/detail/epilepsy.
5. Ken Harris, "The Dangers of Seizures: Why You Need Immediate Treatment," *OSF HealthCare* (blog), August 8, 2018, https://www.osfhealthcare.org/blog/dangers-of-seizures/.
6. Daniel Kablack, "Injuries, Epilepsy End Baltimore Ravens CB Samari Rolle's Ca-

reer," *Bleacher Report*, April 13, 2010, https://bleacherreport.com/articles/377974-injuries-and-epilepsy-end-ravens-cornerback-samari-rolles-career.

7. Tom Pelissero and Ian Rapoport, "Ohio State QB Justin Fields Managing Epilepsy as He Heads into 2021 NFL Draft," *NFL.com*, April 21, 2021, https://www.nfl.com/news/ohio-state-qb-justin-fields-has-confirmed-to-nfl-teams-he-s-managing-epilepsy; "5 NFL Players with Epilepsy," *Epsy Health Blog*, Epsy, February 3, 2022, https://www.epsyhealth.com/seizure-epilepsy-blog/5-nfl-players-with-epilepsy. Other football stars like run-blocker Alan Faneca for the Pittsburgh Steelers, New York Giants' greatest running back ever Tiki Barber, and Jason Snelling for the Atlanta Falcons all suffered from epilepsy, too, and went on to have successful careers in American football.
8. "Drug Resistant Epilepsy," Epilepsy Foundation, last modified October 5, 2020, https://www.epilepsy.com/learn/drug-resistant-epilepsy.
9. Xinzhoung Zhu et al., "Automated Epileptic Seizure Detection in Scalp EEG Based on Spatial-Temporal Complexity," *Complexity* 2017 (2017), https://doi.org/10.1155/2017/5674392.
10. K. Holt, "Researchers Say They Can Predict Epileptic Seizures and Hour in Advance," *Engadget*, Sept. 29, 2020, https://tinyurl.com/8bnwvcjj.
11. Kathy Gilsinan, "The Buddhist and the Neuroscientist," *Atlantic*, July 2015, https://www.theatlantic.com/health/archive/2015/07/dalai-lama-neuroscience-compassion/397706/; Lauren Effron, "Neuroscientist Richie Davidson Says Dalai Lama Gave Him 'a Total Wake-Up Call' that Changed His Research Forever," ABC News, July 27, 2016, https://abcnews.go.com/Health/neuroscientist-richie-davidson-dalai-lama-gave-total-wake/story?id=40859233.
12. "Study of Meditation and Brain Waves in Buddhist Monks Confounds Wisconsin Researchers," BrainTap, April 17, 2018, https://braintap.com/study-of-meditation-and-brain-waves-in-buddhist-monks-confounds-wisconsin-researchers/.
13. Daniel Goleman and Richard Davidson, "How Meditation Changes Your Brain—and Your Life," *Lion's Roar*, May 7, 2018, https://www.lionsroar.com/how-meditation-changes-your-brain-and-your-life/.
14. Antoine Lutz et al., "Long-Term Meditators Self-Induce High-Amplitude Gamma Synchrony During Mental Practice," *Proceedings of the National Academy of Sciences of the United States of America* 101, no. 46 (2004): 16369–16373, https://doi.org/10.1073/pnas.0407401101.
15. Decho Surangsrirat and Apichart Intarapanich, "Analysis of the Meditation Brainwave from Consumer EEG Device," *SoutheastCon 2015* (2015): 1–6, https://doi.org/10.1109/SECON.2015.7133005.
16. Rebecca L. Acabchuk et al., "Measuring Meditation Progress with a Consumer-Grade EEG Device: Caution from a Randomized Controlled Trial," *Mindfulness* 12, no. 1 (2021): 68–81, https://doi.org/10.1007/s12671-020-01497-1; Caroline Stockman, "Can a Technology Teach Meditation? Experiencing the EEG Headband InteraXon Muse as a Meditation Guide," *International Journal of Emerging Technologies in Learning* 15, no. 8 (2020): 83–99; Hugh Hunkin, Daniel L. King, and Ian T. Zajac, "EEG Neurofeedback During Focused Attention Meditation: Effects on State Mindfulness and Meditation Experiences," *Mindfulness* 12 (2021): 841–51, https://doi.org/10.1007/s12671-020-01541-0.
17. Nardin Samuel et al., "Consumer-Grade Electroencephlography Devices as Potential Tools for Early Detection of Brain Tumors," *BMC Medicine* 19, no. 16 (2021), https://doi.org/10.1186/s12916-020-01889-z; Hiroko Ohgaki and Paul Kleihues, "Genetic Pathways to Primary and Secondary Glioblastoma," *American Journal of Pathology* 170, no. 5 (2007): 1445–53, https://doi.org/10.2353/ajpath.2007.070011.

18. M. P. L. Perera and S. R. Liyanage. "Applications and Challenges in Human-Computer Interaction for EEG-Based BCI Systems." *Global Journal of Scientific and Research Publications (GJSRP)* 1, no. 3 (2021): 1–8; Zahra Tabanfar et al., "Brain Tumor Detection Using Electroencephologram Linear and Non-Linear Features," *Iranian Journal of Biomedical Engineering* 10, no. 3 (2016): 211–221, https://dx.doi.org/10.22041/ijbme.2017.72077.1260.
19. iMediSync, *iSyncBrain: Advanced EEG Analysis Platform*, Seoul: iMediSync, Inc., accessed July 12, 2022, https://medicalkorea.micehub-gov.com/home/2021/mdk2021/Files/mdk2021_20210305_144234.pdf.
20. iMediSync, "iMediSync Successfully Debuted AI Mental Care Platform at CES," news release, January 21, 2021, https://ces.vporoom.com/iMediSync/iMediSync-successfully-debuted-AI-mental-care-platform-at-CES-2021.
21. Eric Topol, *The Patient Will See You Now: The Future of Medicine Is in Your Hands* (New York: Basic Books, 2015), 34.
22. Topol, *Patient Will See You.*
23. Cara D. Edwards and Brooke Killian Kim, "The Learned Intermediary Doctrine in the WebMD Era," *DLA Piper,* August 1, 2019, https://www.dlapiper.com/en/us/insights/publications/2019/06/the-learned-intermediary-doctrine-in-the-webmd-era/.
24. Barbara J. Evans, "The Genetic Information Nondiscrimination Act at Age 10: GINA's Controversial Assertion that Data Transparency Protects Privacy and Civil Rights," *William and Mary Law Review* 60, no. 6 (2019): 2017–2109, https://www.ncbi.nlm.nih.gov/pmc/articles/PMC8095822/.
25. *The Farewell,* directed by Lulu Wang (A24, 2019).
26. Leslie J. Blackhall et al., "Bioethics in a Different Tongue: The Case of Truth-Telling," *Journal of Urban Health* 78, no. 1 (2001): 59–71, https://doi.org/10.1093/jurban/78.1.59.; C Y Tse, Alice Chong, and S Y Fok, "Breaking Bad News: A Chinese Perspective," *Palliative Medicine* 17, no. 4 (2003): 339, https://doi.org/10.1191%2F0269216303pm751oa.
27. Hongchun Wang et al., "To Tell or Not: The Chinese Doctors' Dilemma on Disclosure of a Cancer Diagnosis to the Patient," 47 *Iranian Journal of Public Health* 47, no. 11 (Nov 2018): 1773–1774, https://www.ncbi.nlm.nih.gov/pmc/articles/PMC6294856/.
28. Donald Oken, "What to Tell Cancer Patients. A Study of Medical Attitudes," *JAMA* 175 (April 1961): 1120–28, https://pubmed.ncbi.nlm.nih.gov/13730593/; Donna Lu, "The Farewell Explores the Ethics of Lying about a Cancer Diagnosis," *New Scientist,* October 29, 2019, https://www.newscientist.com/article/2221673-the-farewell-explores-the-ethics-of-lying-about-a-cancer-diagnosis/#ixzz6QLC2l25Q.
29. Young Ho Yun et al., "Impact of Awareness of Terminal Illness and Use of Palliative Care or Intensive Care Unit on the Survival of Terminally Ill Patients with Cancer: Prospective Cohort Study," *Journal of Clinical Oncology* 29, no. 18 (June 20, 2011): 2474–80, https://doi.org/10.1200/jco.2010.30.1184; Young Ho Yun et al., "Experiences and Attitudes of Patients with Terminal Cancer and Their Family Caregivers Toward the Disclosure of Terminal Illness," *Journal of Clinical Oncology* 28, no. 11 (April 10, 2010): 1950–57, https://doi.org/10.1200/jco.2009.22.9658; Myung Kyung Lee et al., "Awareness of Incurable Cancer Status and Health-Related Quality of Life Among Advanced Cancer Patients: A Prospective Cohort Study," *Palliative Medicine* 27, no. 2 (2013): 144–54, https://doi.org/10.1177%2F0269216311429042.
30. Miao Wan et al., "The Impact on Quality of Life from Informing Diagnosis in Patients with Cancer: A Systematic Review and Meta-Analysis," *BMC Cancer* 20 (2020): 618, https://doi.org/10.1186/s12885-020-07096-6.
31. Kurt D. Christensen et al., "Associations Between Self-Referral and Health Behavior Responses to Genetic Risk Information," *Genome Medicine* 7, no. 1 (2015): 1–11, https://doi.org/10.1186/s13073-014-0124-0.

32. Ines Testoni et al., "Lack of Truth-Telling in Palliative Care and Its Effects Among Nurses and Nursing Students," *Behavioral Sciences* 10, no. 5 (2020): 88, https://doi.org/10.3390/bs10050088.
33. Erica S. Spatz, Harlan M. Krumholz, and Benjamin W. Moulton, "The New Era of Informed Consent: Getting to a Reasonable Patient Standard Through Shared Decision Making," *JAMA* 315, no. 19 (2016): 2063–64, https://doi.org/10.1001%2Fjama.2016.3070.
34. Richard Fielding and Josephine Hung, "Preferences for Information and Involvement in Decisions During Cancer Care Among a Hong Kong Chinese Population," *Psycho-Oncology* 5, no. 4 (1996): 321–29, https://doi.org/10.1002/(SICI)1099-1611(199612)5:4<321::AID-PON226>3.0.CO;2-K.
35. Tamar Sharon, "Self-Tracking for Health and the Quantified Self: Re-Articulating Autonomy, Solidarity, and Authenticity in an Age of Personalized Healthcare," *Philosophy & Technology* 30, no. 1 (March 2017): 93–121, https://doi.org/10.1007/s13347-016-0215-5.
36. Ewa Grodzinsky and Märta Sund Levander, "History of the Thermometer," *Understanding Fever and Body Temperature: A Cross-Disciplinary Approach to Clinical Practice*, ed. Ewa Grodzinsky and Märta Sund Levander (London: Palgrave Macmillan, 2019): 23–35.
37. Maria da Capua, "Pregnancy Testing Through the Ages: How Lateral Flow Technology Re-Invented the Modern Home Pregnancy Kit," American Association for Clinical Chemistry, March 1, 2020, https://www.aacc.org/cln/cln-industry-insights/2020/pregnancy-testing-through-the-ages.
38. Jesse Olszynko-Gryn, "The Feminist Appropriation of Pregnancy Testing in 1970s Britain," *Women's History Review* 28, no. 6 (2017): 869–894, https://doi.org/10.1080/09612025.2017.1346869.
39. Cari Romm, "Before There Were Home Pregnancy Tests: How Women Found Out They Were Pregnant When They Couldn't Just Pee on a Stick," *The Atlantic*, June 17, 2015, https://www.theatlantic.com/health/archive/2015/06/history-home-pregnancy-test/396077/; "The Thin Blue Line: The History of the Pregnancy Test," Timeline, National Institutes of Health, accessed July 12, 2022, https://history.nih.gov/display/history/Pregnancy+Test+Timeline.
40. Rebecca Kolberg, "A Public Policy Expert Charged Thursday Government Inaction On . . . ," *UPI Archives*, April 6, 1989, https://perma.cc/26DR-UNVU (accessed July 9, 2017).
41. Shelby Baird, "Don't Try This at Home: The FDA's Restrictive Regulation of Home-Testing Devices." *Duke Law Journal* 67 (2017): 404.
42. Public Health England, *HIV Testing and Self-Testing Information Update (November 2015)*, by SJ Westrop et al., London: PHE Publications, 2015, https://assets.publishing.service.gov.uk/government/uploads/system/uploads/attachment_data/file/769460/HIV_Self-Testing_PHE_Position_v13_-_Nov_15_updated.pdf.
43. Shan Juan, "Self-Testing to Boost HIV Battle," *China Daily*, April 21, 2016, https://www.chinadaily.com.cn/china/2016-04/21/content_24711507.htm.
44. Linda Avey, in discussion with the author, May 21, 2021 (transcript on file with author).
45. Linda Avey, in discussion with the author.
46. George J. Annas and Sherman Elias, "23andMe and the FDA," *New England Journal of Medicine* 370, no. 11 (2014): 985, https://doi.org/10.1056/NEJMp1316367.
47. Frank James, "FDA Suggests Consumer Gene Test Needs Agency OK," *NPR*, May 12, 2010, https://www.npr.org/sections/thetwo-way/2010/05/fda_suggests_consumer_gene_tes.html.
48. Alberto Gutierrez to Anne Wojcicki, Letter, U.S. Food and Drug Administration, June 10, 2010, https://www.fda.gov/media/79205/download.

49. 23andMe, "23andMe Takes First Step Toward FDA Clearance," press release, July 30, 2012, https://mediacenter.23andme.com/press-releases/23andme-takes-first-step-toward-fda-clearance.
50. Alberto Gutierrez to Anne Wojcicki, Letter.
51. 23andMe, "23andMe Launches First National TV Campaign," press release, August 5, 2013, https://mediacenter.23andme.com/press-releases/poh_ad_campaign.
52. Alberto Gutierrez to Anne Wojcicki, Letter.
53. Alberto Gutierrez to Anne Wojcicki, Letter.
54. Alberto Gutierrez to Anne Wojcicki, Letter.
55. Alberto Gutierrez to Anne Wojcicki, Letter.
56. Andrew Pollack, "F.D.A. Orders Genetic Testing Firm to Stop Selling DNA Analysis Service," *New York Times*, November 25, 2013, https://www.nytimes.com/2013/11/26/business/fda-demands-a-halt-to-a-dna-test-kits-marketing.html.
57. "23andMe's Updates Regarding FDA's Review," *23andMe Blog*, December 5, 2013, https://blog.23andme.com/news/fda-update/.
58. Scott Hensley, "23andMe Bows to FDA's Demands, Drops Health Claims," NPR, December 6, 2013, https://www.npr.org/sections/health-shots/2013/12/06/249231236/23andme-bows-to-fdas-demands-drops-health-claims.
59. US Food and Drug Administration, "FDA Authorizes, with Special Controls, Direct-to-Consumer Test that Reports Three Mutations in the BRCA Breast Cancer Genes," press release, March 6, 2018, https://www.fda.gov/news-events/press-announcements/fda-authorizes-special-controls-direct-consumer-test-reports-three-mutations-brca-breast-cancer.
60. Louiza Kalokairinou et al., "Legislation of Direct-to-Consumer Genetic Testing in Europe: A Fragmented Regulatory Landscape," *Journal of Community Genetics* 9, no. 2 (2018): 117, 121, https://doi.org/10.1007/s12687-017-0344–2; Rei Fukuda and Fumio Takada, "Legal Regulations on Health-Related Direct-to-Consumer Genetic Testing in 11 Countries," *Kitasato Medical Journal* 48 (2018): 52–59.
61. Pollack, "Stop Selling DNA Analysis."
62. Robert C. Green and Nita A. Farahany, "Regulation: The FDA Is Overcautious on Consumer Genomics," *Nature* 505, no. 7483 (January 2014): 286–87, https://doi.org/10.1038/505286a.
63. Barbara J. Evans, "The Genetic Information Nondiscrimination Act at Age 10: GINA's Controversial Assertion that Data Transparency Protects Privacy and Civil Rights," *William and Mary Law Review* 60, no. 6 (2019): 2017–2109, https://www.ncbi.nlm.nih.gov/pmc/articles/PMC8095822/.
64. Anna Wexler and Robert Thibault, "Mind-reading or Misleading? Assessing Direct-to-Consumer Electroencephalography (EEG) Devices Marketed for Wellness and Their Ethical and Regulatory Implications," *Journal of Cognitive Enhancement* 3, no. 1 (2019): 131–37, https://doi.org/10.1007/s41465-018-0091–2.
65. Caitlin Shure, "Left to Their Own Devices? FDA Policy & Consumer Neurotechnology: Comments on 'General Wellness: Policy for Low-Risk Devices' Draft Guidance Issued 1/20/2015," Comment on US Food and Drug Administration Document ID FDA-2014-N-1039–0002, submitted April 20, 2015, https://www.regulations.gov/comment/FDA-2014-N-1039–0011.
66. Shure, "Left to Their Own Devices?"
67. US Food and Drug Administration, "Sec. 882.1400 Electroencephalograph," *Code of Federal Regulations*, title 21, vol. 8, 21CFR882.1400, last modified March 29, 2022. EEG is "used to measure and record the electrical activity of the patient's brain obtained by placing two or more electrodes on the head."
68. US Food and Drug Administration, "Sec. 890.1375 Diagnostic Electromyograph," *Code of Federal Regulations*, title 21, vol. 8, 21CFR890.1375, last modified March 29,

2022. "A diagnostic electromyograph is a device intended for medical purposes, such as to monitor and display the bioelectric signals produced by muscles, to stimulate peripheral nerves, and to monitor and display the electrical activity produced by nerves, for the diagnosis and prognosis of neuromuscular disease."

69. US Food and Drug Administration Center for Devices and Radiological Health, *General Wellness: Policy for Low-Risk Devices: Guidance for Industry and Food and Drug Administration Staff*, July 29, 2016, updated September 27, 2019, https://www.fda.gov/media/90652/download.
70. FDA, *General Wellness.*
71. FDA.
72. "Cat Ears Moving with Brain Waves Necomimi," accessed August 19. 2022, https://necomimi.shop/en.
73. FocusBand, homepage, accessed July 13, 2022, https://focusband.com.
74. Lewis Gordon, "Brain-Controlled Gaming Exists, Though Ethical Questions Loom over the Tech," *Washington Post*, December 16, 2020, https://www.washingtonpost.com/video-games/2020/12/16/brain-computer-gaming/.
75. "Emotiv Apps," Index.co, accessed July 13, 2022, https://index.co/company/emotiv/apps.
76. "BrainStation," Neuroverse, accessed July 13, 2022, https://www.neuroverseinc.com/brainstation.
77. "Muse," homepage, InteraXon, Inc., accessed July 13, 2022, https://choosemuse.com.
78. Eliza Strickland, "Startup Neurable Unveils the World's First Brain-Controlled VR Game," *IEEE Spectrum*, August 7, 2017, https://spectrum.ieee.org/the-human-os/biomedical/bionics/brainy-startup-neurable-unveils-the-worlds-first-braincontrolled-vr-game.
79. Daniel Johnston, MD, MPH, Comment on US Food and Drug Administration Document ID FDA-2014-N-1039–0002, June 18, 2015, accessed July 29, 2016, https://www.regulations.gov/comment/FDA-2014-N-1039–0006.
80. Alison George, "Think Your Sense of Self Is Located in Your Brain? Think Again," *New Scientist*, December 9, 2020, https://www.newscientist.com/article/mg24833121–500-think-your-sense-of-self-is-located-in-your-brain-think-again/.
81. Christina Starmans and Paul Bloom, "Windows to the Soul: Children and Adults See the Eyes as the Location of the Self," *Cognition* 123, no. 2 (2012): 313–18, https://doi.org/10.1016/j.cognition.2012.02.002.
82. *Satakunnan Markkinapörssi Oy and Satamedia Oy v. Finland*, App. No. 931/13 (2017 ECtHR Grand Chamber).
83. Antoinette Rouvroy and Yves Poullet, "The Right of Informational Self-Determination and the Value of Self-Development: Reassessing the Importance of Privacy for Democracy," *Reinventing Data Protection?*, ed. Serge Gutwirth et al. (Berlin: Springer Science & Business Media, 2009), 45–76.
84. Gabriel Stilman, "The Right to Our Personal Memories: Informational Self-Determination and the Right to Record and Disclose Our Personal Data," *Journal of Ethics and Emerging Technologies* 25, no. 2 (October 2015): 14–24, https://doi.org/10.55613/jeet.v25i2.45.
85. UN Human Rights Council, *Freedom of Opinion and Expression: Report of the Office of the United Nations High Commissioner for Human Rights*, A/HRC/49/38, January 10, 2022, https://documents-dds-ny.un.org/doc/UNDOC/GEN/G22/003/87/PDF/G2200387.pdf.
86. UN Commission on Human Rights Committee, *CCPR General Comment No. 34: Article 19 (Freedoms of Opinion and Expression)*, CCPR/C/GC/34 (September 12, 2011), https://www2.ohchr.org/english/bodies/hrc/docs/gc34.pdf.

87. Stilman, "Our Personal Memories."
88. James P. Evans and Robert C. Green, "Direct to Consumer Genetic Testing: Avoiding a Culture War," *Genetics in Medicine* 11, no. 8 (2009): 568–69, https://doi.org/10.1097/GIM.0b013e3181afbaed.

Chapter 5: Revving Up

1. Martin Dresler et. al., "Hacking the Brain: Dimensions of Cognitive Enhancement," *ACS Chemical Neuroscience* 10, no. 3 (2018): 1137–48, https://doi.org/10.1021/acschemneuro.8b00571.
2. See, for example, David Adam, *The Genius Within* (New York: Pegasus Books, 2018).
3. Larissa J. Maier, Jason A. Ferris, and Adam R. Winstock "Pharmacological Cognitive Enhancement Among Non-ADHD Individuals—A Cross-Sectional Study in 15 Countries," *International Journal of Drug Policy* 58 (August 2018): 104–12, https://doi.org/10.1016/j.drugpo.2018.05.009. The researchers asked a hundred thousand healthy individuals (age sixteen and over) across fifteen different countries whether they had ever used prescription or illegal stimulants to increase their performance at work or while studying. The self-reported rate of use among respondents increased from 4.9 percent in 2015 to 13.7 percent in 2017.
4. Arran Frood, "Use of 'Smart Drugs' on the Rise," *Nature*, July 6, 2018, https://www.scientificamerican.com/article/use-of-ldquo-smart-drugs-rdquo-on-the-rise/.
5. Adrienne Dunn, "Hard Pill to Swallow: Student Adderall Use On the Rise," *State Press* (AZ), September 25, 2018, https://www.statepress.com/article/2018/09/spcommunity-adderall-abuse-on-campuses-student-dealers-and-users.
6. See, for example, Ruairidh M. Battleday and A-K. Brem, "Modafinil for Cognitive Neuroenhancement in Healthy Non-Sleep-Deprived Subjects: A Systematic Review," *European Neuropsychopharmacology* 25, no. 11 (2015): 1865–81, https://doi.org/10.1016/j.euroneuro.2015.07.028; Irena P. Ilieva, Cayce J. Hook, and Martha J. Farah, "Prescription Stimulants' Effects on Healthy Inhibitory Control, Working Memory, and Episodic Memory: A Meta-Analysis," *Journal of Cognitive Neuroscience* 27, no. 6 (June 2015): 1069–89, https://doi.org/10.1162/jocn_a_00776.
7. Battleday and Brem, "Modafinil for Cognitive Neuroenhancement."
8. L-S. Camilla d'Angelo, George Savulich, and Barbara J. Sahakian, "Lifestyle Use of Drugs by Healthy People for Enhancing Cognition, Creativity, Motivation and Pleasure," *British Journal of Pharmacology* 174, no. 19 (2017): 3257–67, https://doi.org/10.1111/bph.13813.
9. Stan B. Floresco and James D. Jentsch, "Pharmacological Enhancement of Memory and Executive Functioning in Laboratory Animals," *Neuropsychopharmacology* 36, no. 1 (2011): 227–50, https://doi.org/10.1038%2Fnpp.2010.158.
10. "Don't Buy into Brain Health Supplements," *Harvard Health Publishing* (blog), March 3, 2022, https://www.health.harvard.edu/mind-and-mood/dont-buy-into-brain-health-supplements.
11. Christina Aungst, "Can Prevagen Really Improve My Memory?" *GoodRx Health*, GoodRx, May 18, 2021, https://www.goodrx.com/well-being/supplements-herbs/prevagen-for-memory-loss-claims; Christian Jarrett, "Do Nootropics Really Work?" *BBC Science Focus*, April 28, 2022, https://www.sciencefocus.com/science/do-nootropics-really-work/.
12. Research and Markets, "Global Cognitive Assessment and Training Market 2017–2021: Aging Population, Increasing Awareness for Brain Fitness & Advancement in Technology Drive the $8.06 Billion Market," *PR Newswire*, April 21, 2017, https://www.prnewswire.com/news-releases/global-cognitive-assessment-and-training-market-2017–2021-aging-population-increasing-awareness-for-brain-fitness—advancement-in-technology-drive-the-806-billion-market—research-and-markets-300443424.html.

13. David Z. Hambrick, "New Evidence Shows Brain Training Games Don't Work," interview by Christie Nicholson, *ZDNet*, May 28, 2012, https://www.zdnet.com/article/qa-new-evidence-shows-brain-training-games-dont-work/.
14. Adrian M. Owen et al., "Putting Brain Training to the Test," *Nature* 465 (2010): 775–78 https://doi.org/10.1038/nature09042; cf. Alexandra B. Morrison and Jason M. Chein, "Does Working Memory Training Work? The Promise and Challenges of Enhancing Cognition by Training Working Memory," *Psychonomic Bulletin & Review* 18, no. 1 (2011): 46–60, https://doi.org/10.3758/s13423-010-0034-0.
15. "A Consensus on the Brain Training Industry from the Scientific Community (Full Statement)," Max Planck Institute for Human Development and Stanford Center on Longevity, October 20, 2014, accessed July 22, 2022, http://longevity.stanford.edu/a-consensus-on-the-brain-training-industry-from-the-scientific-community-2/.
16. Michael Grothaus, "This Is the Only Type of Brain Training That Works, According to Science," *Fast Company*, August 21, 2017, https://www.fastcompany.com/40451692/this-is-the-only-type-of-brain-training-that-works-according-to-science; Tejal M. Shah et al., "Enhancing Cognitive Functioning in Healthy Older Adults: a Systematic Review of the Clinical Significance of Commercially Available Computerized Cognitive Training in Preventing Cognitive Decline," *Neuropsychology Review* 27, no. 1 (2017): 62–80, https://doi.org/10.1007/s11065-016-9338-9. Their conclusion is similarly supported by the largest study to date conducted on brain-training, where 2,832 participants who participated in a set of targeted brain-training exercises were followed for more than a decade. Participants got better at the tasks that they received training in, such as in memory, or other task-related skills.
17. Ranganatha Sitaram et al., "Closed-Loop Brain Training: The Science of Neurofeedback," *Nature Reviews Neuroscience* 18, no. 2 (2017): 86–100, https://doi.org/10.1038/nrn.2016.164.
18. Infiniti Research, *Global Neurofeedback Systems Market 2022–2026* (Elmhurst, STATE: Infiniti Research, 2022).
19. Lucille Nalbach Tournas and Walter G. Johnson, "Elon Musk Wants to Hack Your Brain: How Will the FDA Manage That?" *Slate*, August 5, 2019, https://slate.com/technology/2019/08/elon-musk-neuralink-facebook-brain-computer-interface-fda.html.
20. Konstantinos Mantantzis, Elizabeth A. Maylor, and Friederike Schlaghecken, "Gain Without Pain: Glucose Promotes Cognitive Engagement and Protects Positive Affect in Older Adults," *Psychology and Aging* 33, no. 5 (2018): 789–97, https://psycnet.apa.org/doi/10.1037/pag0000270.
21. Jamie Tully et al., "Estimated Prevalence, Effects, and Potential Risks of Substances Used for Cognitive Enhancement," in *Human Enhancement Drugs*, ed. Katinka van de Ven, Kyle J. D. Mulrooney, and Jim McVeigh (London: Routledge, 2019), 113.
22. Martin Dresler et. al., "Hacking the Brain: Dimensions of Cognitive Enhancement," *ACS Chemical Neuroscience* 10, no. 3 (2018): 1137–48, https://doi.org/10.1021/acschemneuro.8b00571.
23. Laura Smith, "Affidavit in Support of Criminal Complaint," March 11, 2019, https://www.justice.gov/file/1142876/download.
24. Smith, "Affidavit"; Melissa Korn, Jennifer Levitz, and Erin Ailworth, "Federal Prosecutors Charge Dozens in College Admissions Cheating Scheme," *Wall Street Journal*, March 12, 2019, https://www.wsj.com/articles/federal-prosecutors-charge-dozens-in-broad-college-admissions-fraud-scheme-11552403149.
25. Alia Wong, "Why the College Admissions Scandal Is So Absurd," *Atlantic*, March 12, 2019, https://www.theatlantic.com/education/archive/2019/03/college-admissions-scandal-fbi-targets-wealthy-parents/584695/.
26. Libby Nelson, "The Real College Admissions Scandal Is What's Legal," *Vox*, March

12, 2019, https://www.vox.com/2019/3/12/18262037/college-admissions-scandal-felicity-huffman; EJ Dickson, "The College Admissions Scandal Proves the System Is Broken," *Rolling Stone,* March 13, 2019, https://www.rollingstone.com/culture/culture-features/college-admissions-scam-system-broken-807497/.

27. Wong, "Scandal Is So Absurd." *Atlantic,* March 12, 2019, https://www.theatlantic.com/education/archive/2019/03/college-admissions-scandal-fbi-targets-wealthy-parents/584695/.
28. Loretta Tauginienė and Vaidas Jurkevičius, "Ethical and Legal Observations on Contract Cheating Services as an Agreement," *International Journal for Educational Integrity* 13, no. 1 (2017): 1–10, https://doi.org/10.1007/s40979-017-0020-7.
29. Higher Education Standards Panel (HESP), "Overview-Draft Legislation to Tackle Contract Cheating," Australian Department of Education, May 17, 2021, https://www.dese.gov.au/download/4549/overview-draft-legislation-tackle-contract-cheating/6775/document/pdf.
30. Wang Keju, "Court Punishes Organized Exam Cheaters," *China Daily,* August 8, 2018, http://www.chinadaily.com.cn/a/201808/08/WS5b6a446aa310add14f384835.html.
31. "Academic Dishonesty," Student Conduct, Duke University, accessed July 12, 2022, https://studentaffairs.duke.edu/conduct/z-policies/academic-dishonesty.
32. Miranda Katz, "Students Torn Over 'Study Drug' Usage," *Wesleyan Argus* (CT), October 4, 2012, http://wesleyanargus.com/2012/10/04/study-drug-usage-remains-issue-of-heated-debate/.
33. "The Code of Non-Academic Conduct," Wesleyan University, last updated May 2010, accessed July 12, 2022, https://www.wesleyan.edu/studentaffairs/studenthandbook/student-conduct/non-academic-conduct.html.
34. Matt Lamkin, "A Ban on Brain-Boosting Drugs Is Not the Answer," *Chronicle of Higher Education,* February 27, 2011, https://www.chronicle.com/article/a-ban-on-brain-boosting-drugs-is-not-the-answer/.
35. Alan Schwarz, "Attention-Deficit Drugs Face New Campus Rules," *New York Times,* April 30, 2013, https://www.nytimes.com/2013/05/01/us/colleges-tackle-illicit-use-of-adhd-pills.html.
36. Schwarz, "New Campus Rules."
37. Margaret Talbot, "Brain Gain: The Underground World of 'Neuroenhancing' Drugs," *New Yorker,* April 27, 2009, https://www.newyorker.com/magazine/2009/04/27/brain-gain.
38. Philip Hersh, "Marion Jones Shocked Track and Field Fans Two Years Ago When She Said She Would Try for 5 Gold Medals in Sydney. Her Personal Growth and Performances Since Leave Little Doubt That She Has an Excellent Chance," *Chicago Tribune,* September 10, 2000, https://www.chicagotribune.com/news/ct-xpm-2000-09-10-0009100445-story.html.
39. Julia Reed, "Marion Jones: Hail Marion," *Vogue*, December 31, 2000, https://www.vogue.com/article/marion-jones-hail-marion.
40. Reed, "Marion Jones."
41. Associated Press, "Jones Goes Quietly in Athens," ESPN, August 24, 2004, https://www.espn.com/olympics/summer04/trackandfield/news/story?id=1866971.
42. "IOC Formally Strips Marion Jones of Five Sydney Olympic Medals," *New York Times,* December 12, 2007, https://www.nytimes.com/2007/12/12/sports/12iht-olympics12.8712082.html.
43. "The World Anti-Doping Code," World Anti-Doping Agency (WADA), revised January 1, 2021, https://www.wada-ama.org/en/what-we-do/world-anti-doping-code.
44. Julian Savulescu, Bennett Foddy, and Megan Clayton, "Why We Should Allow Performance Enhancing Drugs in Sport," *British Journal of Sports Medicine* 38, no. 6 (2004): 666–70. http://dx.doi.org/10.1136/bjsm.2003.005249.

45. Justin F. Landy, Daniel K. Walco, and Daniel M. Bartels, "What's Wrong with Using Steroids? Exploring Whether and Why People Oppose the Use of Performance Enhancing Drugs," *Journal of Personality and Social Psychology* 113, no. 3 (2017): 377, https://psycnet.apa.org/doi/10.1037/pspa0000089.
46. Savulescu, Foddy, and Clayton, "Why We Should Allow."
47. Amy Daughters, "The Evolution of Football Equipment," *Bleacher Report*, May 16, 2013, https://bleacherreport.com/articles/1642538-the-evolution-of-football-equipment.
48. Savulescu, Foddy, and Clayton, "Why We Should Allow."
49. Leon Watson, "Chess Players Need Checking for Drugs, Scientists Say," *Telegraph* (UK), January 27, 2017, https://www.telegraph.co.uk/news/2017/01/26/scientists-call-crackdown-doping-chess-players-may-taking-performance/; Andreas G. Franke et. al., "Methylphenidate, Modafinil, and Caffeine for Cognitive Enhancement in Chess: A Double-Blind, Randomised Controlled Trial," *European Neuropsychopharmacology* 27, no. 3 (2017): 248–60, https://doi.org/10.1016/j.euroneuro.2017.01.006.
50. Craig L. Carr, "Coercion and Freedom," *American Philosophical Quarterly* 25, no. 1 (January 1988): 59–67, https://www.jstor.org/stable/20014223.
51. Kimberly J. Schelle et al., "Attitudes Toward Pharmacological Cognitive Enhancement—a Review," *Frontiers in Systems Neuroscience* 8 (2014): 53, https://doi.org/10.3389/fnsys.2014.00053.
52. *Anderson v. Khanna*, 913 N.W.2d 526 (Iowa 2018).
53. See, for example, Mirko D. Garasic and Andrea Lavazza, "Moral and Social Reasons to Acknowledge the Use of Cognitive Enhancers in Competitive-Selective Contexts," *BMC Medical Ethics* 17, no. 1 (2016): 1–12, https://www.ncbi.nlm.nih.gov/pmc/articles/PMC4812634/; B. Sonny Bal and Theodore J. Chomma, "What to Disclose? Revisiting Informed Consent," *Clinical Orthopaedics and Related Research* 470, no. 5 (2012): 1346–56, https://www.ncbi.nlm.nih.gov/pmc/articles/PMC3314757/; Jerry Menikoff, *Law and Bioethics: An Introduction* (Washington, DC: Georgetown University Press, 2002); Johnson v Kokemoor, 545 NW2d 495 (Wis. 1996).

Chapter 6: Braking the Brain

1. Lesley Stahl, *60 Minutes*, "The Memory Pill," produced by Shari Finkelstein (New York: Columbia Broadcasting System, 2007).
2. Alain Brunet et al., "Reduction of PTSD Symptoms with Re-Activation Propranolol Therapy: A Randomized Controlled Study," *American Journal of Psychiatry* 175, no. 5 (May 2018): 427–33, https://doi.org/10.1176/appi.ajp.2017.17050481.
3. Frederico Rotondo et al., "Lack of Effect of Propranolol on the Reconsolidation of Conditioned Fear Memory Due to a Failure to Engage Memory Destabilisation," *Neuroscience* 480 (2022): 9–18, https://doi.org/10.1016/j.neuroscience.2021.11.008; Sanket B. Raut et al., "Effects of Propranolol on the Modification of Trauma Memory Reconsolidation in PTSD Patients: A Systemic Review and Meta-Analysis," *Journal of Psychiatric Research* 150 (June 2022): 246–56, https://doi.org/10.1016/j.jpsychires.2022.03.045.
4. Daniel Kolitz, "Will We Ever Be Able to Edit or Delete Memories," *Gizmodo*, August 20, 2021.
5. Aurelio Cortese et al., "The DecNef Collection, fMRI Data from Closed-Loop Decoded Neurofeedback Experiments," *Scientific Data* 8, no. 1 (2021): 65, https://www.nature.com/articles/s41597-021-00845–7.
6. Michal Ramot and Alex Martin, "Closed-Loop Neuromodulation for Studying Spontaneous Activity and Causality," *Trends in Cognitive Sciences* 26, no. 4 (April 2022): 290–99, https://doi.org/10.1016/j.tics.2022.01.008.
7. Ai Koizuimi and Mitsuo Kawato, "Chapter 10—Implicit Decoded Neurofeedback

Training as a Clinical Tool," in *fMRI Neurofeedback*, ed. Michelle Hampson (Cambridge, MA: Academic Press, 2021), 239–47.

8. Marlene Oscar-Berman and Ksenija Marinković, "Alcohol: Effects on Neurobehavioral Functions and the Brain," *Neuropsychology Review* 17, no. 3 (2007): 239–57, https://doi.org/10.1007/s11065-007-9038-6.
9. Antonia Abbey, Mary Jo Smith, and Richard O. Scott, "The Relationship Between Reasons for Drinking Alcohol and Alcohol Consumption: An Interactional Approach," *Addictive Behaviors* 18, no. 6 (November–December 1993): 659–70, https://doi.org/10.1016%2F0306–4603(93)90019–6; Emmanuel Kuntsche et al., "Who Drinks and Why? A Review of Socio-Demographic, Personality, and Contextual Issues Behind the Drinking Motives in Young People," *Addictive Behaviors* 31, no. 10 (2006): 1844–57, https://doi.org/10.1016/j.addbeh.2005.12.028.
10. Floyd W. Tomkins "Prohibition," *The Annals of the American Academy of Political and Social Science* 109, (September 1923): 15–25, https://www.jstor.org/stable/1014989.
11. "GMA: Lauren Hutton Describes Motorcycle Crash," ABC News, March 1, 2001, https://abcnews.go.com/GMA/story?id=127213&page=1.
12. Bethan Holt, "'There Are So Many Ways of Being Beautiful': Lauren Hutton on Ageing Naturally," *Sydney Morning Herald*, March 7, 2020, https://www.smh.com.au/lifestyle/fashion/there-are-so-many-ways-of-being-beautiful-lauren-hutton-on-ageing-naturally-20200304-p546tn.html.
13. Staci Sturrock, "Hutton Happy Just to Be Alive," *Deseret News*, March 26, 2001, https://www.deseret.com/2001/3/26/19576698/hutton-happy-just-to-be-alive.
14. Rebecca Y. Du et al., "Primary Prevention of Road Traffic Accident-Related Traumatic Brain Injuries in Younger Populations: A Systemic Review of Helmet Legislation," *Journal of Neurosurgery: Pediatrics* 25, no. 4 (2020): 361–74, https://doi.org/10.3171/2019.10.PEDS19377; Michael C. Dewan et al., "Estimating the Global Incidence of Traumatic Brain Injury," *Journal of Neurosurgery* 130, no. 4 (2018): 1080–97, https://doi.org/10.3171/2017.10.JNS17352.
15. "Frequently Asked Questions," Brain Trauma Foundation, accessed July 13, 2022, https://www.braintrauma.org/faq.
16. Du et al., "Accident-Related Traumatic Brain Injuries." Nearly every study on the issue, a systematic Cochrane review, and several meta-analyses show the utility of helmets in decreasing motorcycle-related injuries and deaths. One group of researchers looked at the eighteen different studies of the effectiveness of helmet legislation on motorcycle head injuries and fatalities and found that seventeen of eighteen studies found a positive association between mandatory helmet legislation and a reduction in motorcycle-related injuries or fatalities. Nine of the ten studies focused on the impact of helmet laws on head injuries found a substantial reduction in head injuries related to helmet legislation. And six of the eight studies focused on fatalities found that helmet legislation substantially reduced motorcycle-accident-related fatalities. Only one of the eighteen studies on the issue found no positive association between helmet legislation and reducing the number of injuries or fatalities resulting from motorcycle crashes.
17. Marian Moser Jones and Richard Bayer, "Paternalism & Its Discontents: Motorcycle Helmet Laws, Libertarian Values, and Public Health," *American Journal of Public Health* 97, no. 2 (February 2007): 208–17. In six of the twenty-seven states that require minors-only helmet laws, adult riders must carry at least $10,000 in insurance coverage or wear helmets during their first year of riding.
18. Jacob Lepard et al., "Differences in Outcomes of Mandatory Motorcycle Helmet Legislation by Country Income Level: A Systemic Review and Meta-Analysis," *PLoS Medicine* 18, no. 9 (2021): e1003795, https://doi.org/10.1371/journal.pmed.1003795.
19. Jones and Bayer, "Paternalism & Its Discontents."

20. See, for example, *Simon v. Sargent*, 346 F.Supp. 277 (D. Mass. 1972), *affirmed* 93 S.Ct. 463 (1972); Jones and Bayer, "Paternalism & Its Discontents."
21. *Simon v. Sargent.*
22. Jones and Bayer, "Paternalism & Its Discontents."
23. Kim Treiger-Bar-Am, "In Defense of Autonomy: An Ethic of Care," *NYU Journal of Law and Liberty* 3 (2008): 548, 561.
24. Gerald Dworkin, *The Theory and Practice of Autonomy* (Cambridge: Cambridge University Press, 1988).
25. Dr. Kim Trieger-Bar-Am, *In Defense of Autonomy: An Ethic of Care,* 561.
26. Candace Cummins Gauthier, "Philosophical Foundations of Respect for Autonomy," *Kennedy Institute of Ethics Journal* 3, no. 1 (1993): 30, https://doi.org/10.1353/ken.0.0103.
27. While the principal of self-determination has over time been understood as "the right of peoples and nations to self-determination," I am arguing for an international human right to self-determination over one's own body and mental experiences as undergirding other existing rights. See e.g. Henkin, Louis, Sarah H. Cleveland, Laurence R. Helfer, Gerald L. Neuman, and Diane F. Orenlicher, *Human Rights: Second Edition* (New York: Foundation Press, 2009), 368–70.
28. *Jehovah's Witnesses of Moscow v. Russian Federation*, App. No. 302/02, ECHR (June 10, 2010).
29. Jill Marshall, *Human Rights Law and Personal Identity* (London: Routledge, 2014), 36.
30. John Stuart Mill, *On Liberty* (London: John W. Parker and Son, 1859), 276.
31. Mill, *On Liberty*, 281.
32. Barbara Secker, "The Appearance of Kant's Deontology in Contemporary Kantianism: Concepts of Patient Autonomy in Bioethics," *Journal of Medicine and Philosophy* 24, no. 1 (1999): 43–66, https://doi.org/10.1076/jmep.24.1.43.2544.
33. Kim Treiger-Bar-Am, "In Defense of Autonomy: An Ethic of Care," *NYU Journal of Law and Liberty* 3 (2008): 548, 561.
34. Mill, *On Liberty*, 82.
35. US Food and Drug Administration, "Timeline of Selected FDA Activities and Significant Events Addressing Opioid Misuse and Abuse," last modified June 28, 2022, https://www.fda.gov/drugs/information-drug-class/timeline-selected-fda-activities-and-significant-events-addressing-opioid-misuse-and-abuse.
36. U.S. Department of Health and Human Services, "HHS Acting Secretary Declares Public Health Emergency to Address National Opioid Crisis," press release, October 26, 2017, accessed December 31, 2020, https://public3.pagefreezer.com/browse/HHS.gov/31–12–2020T08:51/https://www.hhs.gov/about/news/2017/10/26/hhs-acting-secretary-declares-public-health-emergency-address-national-opioid-crisis.html.
37. World Health Organization, "Opioid Overdose," fact sheet, August 4, 2021, https://www.who.int/news-room/fact-sheets/detail/opioid-overdose.
38. Paula Alejandro Navarro et al., "Safety and Feasibility of Nuceleus Accumbens Surgery for Drug Addiction: A Systematic Review," *Neuromodulation: Technology at the Neural Interface* 25, no. 2 (2022): 171, https://doi.org/10.1111/ner.13348.
39. "Opioid Overdose Crisis," NIH National Institute on Drug Abuse, last modified March 11, 2021, https://www.drugabuse.gov/drug-topics/opioids/opioid-overdose-crisis.
40. Jeff Nesbit, "We Have Lost the War on Drugs," *U.S. News & World Report,* December 21, 2015, https://www.usnews.com/news/blogs/at-the-edge/articles/2015–12–21/the-war-on-drugs-is-over-and-we-lost.
41. Jennifer Chesebro, "Advocating for Adequate Pain Relief During the Opioid Epidemic," *Nursing* 49, no. 12 (Dec 2019): 64–7, https://doi.org/10.1097/01.NURSE.0000604756.77748.c4.

42. Ausaf Bari et al., "Neuromodulation for Substance Addiction in Human Subjects: A Review," *Neuroscience & Biobehavioral Reviews* 95 (Dec 2018): 34, https://doi.org/10.1016/j.neubiorev.2018.09.013.
43. N. Dafny & G.C. Rosenfeld, "Neurobiology of Drug Abuse," in *Conn's Translational Neuroscience,* ed P. Michael Conn (Cambridge, MA: Academic Press, 2017), 715–722.
44. Nan Li et al., "Nucleus Accumbens Surgery for Addiction," *World Neurosurgery* 80, no. 3–4 (Sep-Oct 2013): S28.E9-S28.E19, https://doi.org/10.1016/j.wneu.2012.10.007.; Benjamin D. Greenberg, Scott L. Rauch and Suzanne N. Haber, "Invasive Circuitry-Based Neurotherapuetics: Stereotactic Ablation and Deep Brain Stimulation for OCD," *Neurpsychopharmacology* 35, no. 1 (Jan 2010), 317–336, https://doi.org/10.1038%2Fnpp.2009.128. Although ablation techniques are sometimes used in the United States to remove brain tumors, ablations of actual brain tissue are exceedingly rare. Only in cases of extreme neuropsychiatric disorders, such as obsessive-compulsive disorder and intractable depression, where traditional treatment has failed, is stereotactic ablation considered appropriate. Requests for the surgery must be approved by a multidisciplinary committee and are regularly rejected.
45. Nan Li et al., "Nucleus Accumbens Surgery for Addiction."
46. Caleb Hellerman, "Experimental Brain Surgery May Help Some People Overcome Drug Addiction," *CNN Health,* February 15, 2022, https://www.cnn.com/2022/02/15/health/drug-addiction-deep-brain-stimulation/index.html.
47. Shuo Ma et al., "Neurosurgical Treatment for Addiction: Lessons from an Untold Story in China and a Path Forward," *National Science Review* 7, no. 3 (March 2020): 702–12, https://doi.org/10.1093/nsr/nwz207; Ausaf Bari et al., "Neuromodulation for Substance Addiction in Human Subjects: A Review," *Neuroscience & Biobehavioral Reviews* 95 (December 2018): 33, https://doi.org/10.1016/j.neubiorev.2018.09.013; and Caleb Hellerman, "Experimental Brain Surgery May Help Some People Overcome Drug Addiction," *CNN Health,* February 15, 2022, https://www.cnn.com/2022/02/15/health/drug-addiction-deep-brain-stimulation/index.html.
48. Paula Alejandro Navarro et al., "Safety and Feasibility of Nuceleus Accumbens Surgery for Drug Addiction: A Systematic Review," *Neuromodulation: Technology at the Neural Interface* 25, no. 2 (2022): 171, https://doi.org/10.1111/ner.13348.; Ausaf Bari et al., "Neuromodulation for Substance Addiction in Human Subjects: A Review."
49. Caleb Hellerman, "Experimental Brain Surgery May Help Some People Overcome Drug Addiction," *CNN Health,* February 15, 2022, https://www.cnn.com/2022/02/15/health/drug-addiction-deep-brain-stimulation/index.html.
50. Elizabeth I. Martin et al., "The Neurobiology of Anxiety Disorders: Brain Imaging, Genetics, and Psychoneuroendocrinology," *Psychiatric Clinics* 32, no. 3 (Sep 2009): 549–575, https://doi.org/10.1016%2Fj.psc.2009.05.004.
51. Chasity Shalon Norris, "Psychopathy and Gender of Serial Killers: A Comparison Using the PCL-R" (master's thesis, East Tennessee State University, 2011), https://dc.etsu.edu/etd/1340/.
52. R. James R. Blair, "The Amygdala and Ventromedial Prefrontal Cortex: Functional Contributions and Dysfunction in Psychopathy," *Philosophical Transactions of the Royal Society B: Biological Sciences* 363, no. 1503 (2008): 2557–2565, https://doi.org/10.1098/rstb.2008.0027.
53. Armineh Zohrabian and Tomas J. Philipson, "External Costs of Risky Health Behaviors Associated with Leading Actual Causes of Death in the U.S.: A Review of the Evidence and Implications for Future Research," *International Journal of Environmental Research and Public Health* 7, no. 6 (2010): 2460–2472, https://doi.org/10.3390/ijerph7062460.

54. Zohrabian and Philipson, "External Costs of Risky Health Behaviors."
55. John Cawley and Christopher J. Ruhm, "The Economics of Risky Health Behaviors," *Handbook of Health Economics* vol .2, eds. Mark V. Pauley, Thomas C. McGuire, and Pedro P. Barros, (Amsterdam: Elsevier, 2012): 95–199.
56. W. Kip Viscusi and Ted Gayer, "Resisting Abuses of Benefit-Cost Analysis," *National Affairs* (Spring 2016), https://www.nationalaffairs.com/publications/detail/resisting-abuses-of-benefit-cost-analysis.
57. Mill, *On Liberty*, 82.
58. Mill, *On Liberty*, 84–6.
59. Mill, *On Liberty*, 101–102.
60. Mill, *On Liberty*, 64.

Chapter 7: Mental Manipulation

1. Baris Korkmaz, "Theory of Mind and Neurodevelopmental Disorders of Childhood," *Pediatric Research* 69, no. 8 (2011): 101–08, https://doi.org/10.1203/PDR.0b013e318212c177.
2. Kendra Cherry, "How the Theory of Mind Helps Us Understand Others," *Verywell Mind*, Dotdash Media, last updated July 4, 2021, https://www.verywellmind.com/theory-of-mind-4176826.
3. Anna R. McAlister and T. Bettina Cornwell, "Preschool Children's Persuasion Knowledge: The Contribution of Theory of Mind," *Journal of Public Policy & Marketing* 28, no. 2 (2009): 175–185, https://doi.org/10.1509/jppm.28.2.175.
4. Baris Korkmaz, "Theory of Mind and Neurodevelopmental Disorders of Childhood," *Pediatric Research* 69, no. 8 (2011): 101–108, https://doi.org/10.1203/PDR.0b013e318212c177.
5. McAlister and Cornwell, "Preschool Children's Persuasion Knowledge: The Contribution of Theory of Mind."
6. "Legilimency," Harry Potter Wiki, last modified May 22, 2022, https://harrypotter.fandom.com/wiki/Legilimency.
7. Kevin Randall, "Rise of Neurocinema: How Hollywood Studios Harness Your Brainwaves to Win Oscars," *Fast Company*, February 25, 2011, https://www.fastcompany.com/1731055/rise-neurocinema-how-hollywood-studios-harness-your-brainwaves-win-oscars.
8. Randall, "Rise of Neurocinema."
9. "The Relationship Between Brain Waves and Blockbuster Movies," Paramount, January 16, 2018, https://www.paramount.com/news/partner-solutions/neuroscience-and-movie-trailers.
10. Sheila Marikar, "'Avatar' Eclipses 'Titanic' to Become Top Box Office Earner," *ABC News*, January 26, 2010, https://abcnews.go.com/Entertainment/Oscars/avatar-titanic-reasons-mega-movie-hits/story?id=9628723.
11. "The Relationship Between Brain Waves and Blockbuster Movies," Paramount.
12. Statista Research Department, "U.S. Digital Advertising Industry-Statistics & Facts," Statista, May 4, 2022, https://www.statista.com/topics/1176/online-advertising/#dossierKeyfigures.
13. Michela Balconi and Martina Sansone, "Neuroscience and Consumer Behavior: Where to Now?," *Frontiers in Psychology* 12 (2021): 705850, https://doi.org/10.3389/fpsyg.2021.705850.
14. Vishali Khurana et al., "A Survey on Neuromarketing Using EEG Signals," *IEEE Transactions on Cognitive and Developmental Systems* 13, no. 4 (2021): 732–49.
15. Eben Harrell, "Neuromarketing: What you Need to Know," *Harvard Business Review* January 23, 2019, https://hbr.org/2019/01/neuromarketing-what-you-need-to-know.

16. Samuel M. McClure et al., "Neural Correlates of Behavioral Preference for Culturally Familiar Drinks," *Neuron* 44, no. 2 (2004): 379–387, https://doi.org/10.1016/j.neuron.2004.09.019.
17. Hilke Plassman et al., "Marketing Actions Can Modulate Neural Representations of Experienced Pleasantness," *Proceedings of the National Academy of Sciences* 105, no. 3 (2008): 1050–54, https://doi.org/10.1073/pnas.0706929105.
18. Fotis P. Kalaganis et al., "Unlocking the Subconscious Consumer Bias: A Survey on the Past, Present, and Future of Hybrid EEG Schemes in Neuromarketing," *Frontiers in Neuroergonomics* 2 (2021): 672982, https://doi.org/10.3389/fnrgo.2021.672982.
19. Harrell, "Neuromarketing: What you Need to Know."
20. Balconi and Sansone, "Neuroscience and Consumer Behavior."
21. Harrell, "Neuromarketing: What you Need to Know."
22. Balconi and Sansone, "Neuroscience and Consumer Behavior,"; Kalaganis et al., "Unlocking the Subconscious Consumer Bias."
23. Harrell, "Neuromarketing: What you Need to Know."
24. Mordor Intelligence, *Global Neuromarketing Market—Growth, Trends, COVID-19 Impact, and Forecasts (2022–2027)*, accessed July 13, 2022, https://www.mordorintelligence.com/industry-reports/neuromarketing-market.
25. Vishali Khurana et al., *A Survey on Neuromarketing Using EEG Signals*, 13 IEEE Transactions on Cognitive and Developmental Systems 732 (December 2021).
26. Anya Kamenetz, "War Displaced Two-thirds of Ukraine's Children. Keeping Them Safe Isn't Easy," *NPR*, June 9, 2022, https://www.npr.org/2022/05/29/1101973267/an-estimated-two-thirds-of-ukrainian-children-have-had-to-leave-their-homes.
27. Osnat Lubrani, "The War Has Caused the Fastest and Largest Displacement of People in Europe Since World War II," United Nations Ukraine, March 22, 2022, https://ukraine.un.org/en/175836-war-has-caused-fastest-and-largest-displacement-people-europe-world-war-ii.
28. Charles Duhigg, "Why Don't You Donate for Syrian Refugees? Blame Bad Marketing," *New York Times*, June 14, 2017, https://www.nytimes.com/2017/06/14/business/media/marketing-charity-water-syria.html.
29. Ana C. Martinez-Levy et al., "Message Framing, Non-Conscious Perception and Effectiveness in Non-Profit Advertising. Contribution by Neuromarketing Research," *International Review on Public and Nonprofit Marketing* 19, no. 1 (2022): 53–75, https://doi.org/10.1007/s12208-021-00289-0.
30. C. Luna-Nevarez, "Neuromarketing, Ethics and Regulation: An Exploratory Analysis of Consumer Opinions and Sentiment on Blogs and Social Media," *Journal of Consumer Policy* 44, no. 4 (2021): 559–583, https://doi.org/10.1007/s10603-021-09496-y.
31. "NMSBA Code of Ethics," https://nmsba.com/neuromarketing-companies/code-of-ethics.
32. Emily R. Murphy et al., "Neuroethics of Neuromarketing," *Journal of Consumer Behaviour: An International Research Review* 7, no. 4-5 (2008): 293–302., https://doi.org/10.1002/cb.252.
33. C. Luna-Nevarez, "Neuromarketing, Ethics and Regulation."
34. Harrell, "Neuromarketing: What you Need to Know."
35. Casey Newton, "The Person Behind the Like Button Says Software is Wasting Our Time," *The Verge*, March 28, 2018, https://www.theverge.com/2018/3/28/17172404/justin-rosenstein-asana-social-media-facebook-timeline-gantt.
36. Paul Lewis, "'Our Minds Can be Hijacked': The Tech Insiders Who Fear a Smartphone Dystopia," *The Guardian*, October 6, 2017, https://www.theguardian.com/technology/2017/oct/05/smartphone-addiction-silicon-valley-dystopia.
37. Verto Analytics, "Average Unlocks per Day Among Smartphone Users in the

United States as of August 2018, by Generation," Chart, April 24, 2019, https://www.statista.com/statistics/1050339/average-unlocks-per-day-us-smartphone-users/.

38. "Cell Phone Addiction: The Statistics of Gadget Dependency," Articles, King University Online, July 27, 2017, https://online.king.edu/news/cell-phone-addiction/.
39. John Brandon, "These Updated Stats About How Often You Use Your Phone Will Humble You," *Inc.*, November 19, 2019, https://www.inc.com/john-brandon/these-updated-stats-about-how-often-we-use-our-phones-will-humble-you.html.
40. Adrian F. Ward et al., "Brain Drain: The Mere Presence of One's Own Smartphone Reduces Available Cognitive Capacity," *Journal of the Association for Consumer Research* 2, no. 2 (Apr 2017): 140–154, https://doi.org/10.1086/691462.
41. Lewis, " 'Our Minds Can be Hijacked.' "
42. Matthew B. Lawrence, "Addiction and Liberty," *Cornell Law Review* 108, Forthcoming 2023, https://dx.doi.org/10.2139/ssrn.4113570.
43. Lawrence, "Addiction and Liberty."
44. Lewis, " 'Our Minds Can be Hijacked.' "
45. Lewis, " 'Our Minds Can be Hijacked.' "
46. "How to Build Habit-Forming Products," https://www.nirandfar.com/hooked-workshop/.
47. Lewis, " 'Our Minds Can be Hijacked.' "
48. Lewis, " 'Our Minds Can be Hijacked.' ""
49. Lewis, " 'Our Minds Can be Hijacked.' "
50. Orge Castellano, "Social Media Giants are Hacking Your Brain—This is How," *Medium*, December 18, 2017, https://orge.medium.com/your-brain-is-being-hacked-by-social-media-584ac1d2083c.
51. Joseph Firth et al., "The 'Online Brain': How the Internet May be Changing Our Cognition," *World Psychiatry* 18, no. 2 (Jun 2019): 119–129, https://doi.org/10.1002/wps.20617.
52. Daniel W. Belsky et al., "Polygenic Risk and the Developmental Progression to Heavy, Persistent Smoking and Nicotine Dependence: Evidence from a 4-Decade Longitudinal Study," *JAMA Psychiatry* 70, no. 5 (2013): 534–542, doi:10.1001/jamapsychiatry.2013.736.
53. Center for Humane Technology, "Social Media and the Brain: Why is Persuasive Technology so Hard to Resist?" Updated August 17, 2021, accessed July 14, 2022, https://www.humanetech.com/youth/social-media-and-the-brain.; Daniel Susser, Beate Roessler, and Helen Nissenbaum, "Online Manipulation: Hidden Influences in a Digital World," *Georgetown Law Technology Review* 4 (2019): 1–45, https://heinonline.org/HOL/P?h=hein.journals/gtltr4&i=13.
54. Lewis, " 'Our Minds Can be Hijacked.' "
55. Tristan Harris, "How Technology is Hijacking Your Mind—From a Magician and Google Design Ethicist," *Medium*, May 18, 2016, https://medium.com/thrive-global/how-technology-hijacks-peoples-minds-from-a-magician-and-google-s-design-ethicist-56d62ef5edf3.
56. Castellano, "Social Media Giants are Hacking Your Brain—This is How."
57. Daniel Susser, Beate Roessler, and Helen Nissenbaum, "Online Manipulation: Hidden Influences in a Digital World," *Georgetown Law Technology Review* 4 (2019): 21, https://heinonline.org/HOL/P?h=hein.journals/gtltr4&i=13.
58. Deena Skolnick Weisberg et al., "The Seductive Allure of Neuroscience Explanations," *Journal of Cognitive Neuroscience*, 20, no. 3 (Mar 2008): 470–477, http://dx.doi.org/10.1162/jocn.2008.20040.
59. See, e.g., David P. McCabe and Alan D. Castel, "Seeing is Believing: The Effect of

Brain Images on Judgements of Scientific Reasoning," *Cognition* 107, no. 1, (April 2008): 343–352, https://doi.org/10.1016/j.cognition.2007.07.017.

60. Deena Skolnick Weisberg, Jordan C. V. Taylor, and Emily J. Hopkins, "Deconstructing the Seductive Allure of Neuroscience Explanations," *Judgment and Decision-Making* 10, no. 5 (2015): 432–433, retrieved from https://repository.upenn.edu/neuroethics_pubs/132.
61. Deena Skolnik Weisberg, Emily J. Hopkins and Jordan C. V. Taylor, *People's Explanatory Preferences for Scientific Phenomena, Cognitive Research: Principles and Implications* 3, no. 1 (2018): 44, https://doi.org/10.1186/s41235-018-0135-2.
62. Gordon Pennycook and David G. Rand, "Lazy, Not Biased: Susceptibility to Partisan Fake News is Better Explained by Lack of Reasoning Than by Motivated Reasoning, " *Cognition* 188 (2019): 39–50, https://doi.org/10.1016/j.cognition.2018.06.011/.
63. Katy Steinmetz, "How Your Brain Tricks You Into Believing Fake News," *Time,* August 9, 2018, https://time.com/5362183/the-real-fake-news-crisis/.
64. Kristen Weir, "Why We Fall for Fake News: Hijacked Thinking or Laziness?," American Psychological Assocation (February 11, 2020), https://www.apa.org/news/apa/2020/fake-news.
65. Steinmetz, "How Your Brain Tricks You Into Believing Fake News."
66. Dina ElBoghdady, "Market Quavers After Fake AP Tweet Says Obama Was Hurt in White House Explosions," *Washington Post,* April 23, 2013, https://www.washingtonpost.com/business/economy/market-quavers-after-fake-ap-tweet-says-obama-was-hurt-in-white-house-explosions/2013/04/23/d96d2dc6-ac4d-11e2-a8b9–2a63d75b5459_story.html.
67. Soroush Vosoughi, Deb Roy, and Sinan Aral, "The Spread of True and False News Online," *Science* 359, no. 6380 (2018): 1146–1151, https://doi.org/10.1126/science.aap9559.
68. Bryan Strange et al., "Information Theory, Novelty, and Hippocampal Responses: Unpredicted or Unpredictable," *Neural Networks* 18, no. 3 (Apr 2005): 225–230, https://doi.org/10.1016/j.neunet.2004.12.004.
69. Vosoughi, Roy, and Aral, "The Spread of True and False News Online."
70. CCDH, *The Disinformation Dozen: Why Platforms Must Act on Twelve Leading Online Anti-Vaxxers* (2021), https://counterhate.com/wp-content/uploads/2022/05/210324-The-Disinformation-Dozen.pdf.
71. Shannon Bond, "Just 12 People are Behind Most Vaccine Hoaxes on Social Media, Research Shows," *NPR* (May 14, 2021), available at https://www.npr.org/2021/05/13/996570855/disinformation-dozen-test-facebooks-twitters-ability-to-curb-vaccine-hoaxes.
72. Center for Countering Digital Hate, *The Disinformation Dozen: Why Platforms Must Act on Twelve Leading Online Anti-Vaxxers,* March 24, 2021, https://counterhate.com/wp-content/uploads/2022/05/210324-The-Disinformation-Dozen.pdf.
73. The Decision Lab, "Why Do We Believe Misinformation More Easily When It's Repeated Many Times?," accessed July 14, 2022,https://thedecisionlab.com/biases/illusory-truth-effect.
74. Simon McCarthy-Jones, "Freedom of Thought is Under Attack—Here's How to Save Your Mind," *The Conversation,* October 21, 2019, https://theconversation.com/freedom-of-thought-is-under-attack-heres-how-to-save-your-mind-124379.
75. B. P. Vermeulen, "Freedom of Thought, Conscience and Religion (Article 9)," *Theory and Practice of the European Convention on Human Rights,* eds. Pieter van Dijk et al., 4th ed. (Cambridge: Intersentia Publishers, 2006): 751–771; Ahmed Shaheed, *Thematic Report on The Freedom of Thought: Advance Unedited Version,* submitted to the seventy-sixth session of the UN General Assembly, A/76/380, (October 5, 2021).

76. William R. Clark and Michael Grunstein, *Are We Hardwired?: The Role of Genes in Human Behavior* (New York: Oxford University Press, 2000), 265; John L. Hill, "Note: Freedom, Determinism, and The Externalization of Responsibility in the Law: A Philosophical Analysis," *Georgetown Law Journal* 76 (1988): 2045.
77. Nita A. Farahany, "A Neurological Foundation for Freedom," *Stanford Technology Law Review* (2012): 4, https://heinonline.org/HOL/P?h=hein.journals/stantlr2012&i=127.
78. Harry G. Frankfurt, "Freedom of Will and Concept of a Person," *Journal of Philosophy* 68, no. 1 (1971): 5, 7, https://doi.org/10.2307/2024717.
79. Frankfurt, "Freedom of Will and Concept of a Person," 13.
80. Frankfurt, "Freedom of Will and Concept of a Person," 13.
81. Gerald Dworkin, "Autonomy and Behavioral Control," *The Hastings Center Report* 6, no. 1 (Feb 1976): 23–28, https://doi.org/10.2307/3560358.
82. Daniel Susser, Beate Roessler, and Helen Nissenbaum, "Online Manipulation: Hidden Influences in a Digital World," *Georgetown Law Technology Review* 4 (2019): 13, https://heinonline.org/HOL/P?h=hein.journals/gtltr4&i=13.
83. Susser, Roessler, and Nissenbaum, "Online Manipulation," 16, 17.
84. Andrea Lavazza, "Freedom of Thought and Mental Integrity: The Moral Requirements for Any Neural Prosthesis," *Frontiers in Neuroscience* 12 (2018): 82, https://doi.org/10.3389/fnins.2018.00082.
85. Marcello Ienca and Roberto Andorno, "Towards New Human Rights in the Age of Neuroscience and Neurotechnology." *Life Sciences, Society and Policy* 13, no. 1 (2017):5, https://doi.org/10.1186/s40504-017-0050–1.
86. Daniel Susser et al., *Online Manipulation: Hidden Influences in a Digital World*, 38.
87. John A. Bargh, "The Hidden Life of the Consumer Mind," *Consumer Psychology Review* 5, no. 1 (Jan 2022): 3–18, https://doi.org/10.1002/arcp.1075.
88. Bargh, "The Hidden Life of the Consumer Mind," 3018.
89. Bargh, "The Hidden Life of the Consumer Mind," 3018.
90. Gráinne M. Fitzimons, Tanya L. Chartrand, and Gavan J. Fitzsimons, "Automatic Effects of Brand Exposure on Motivated Behavior: How Apple Makes You 'Think Different,'" *Journal of Consumer Research* 35, no. 1 (Jun 2008): 21–35, https://doi.org/10.1086/527269.
91. Bargh, "The Hidden Life of the Consumer Mind," 3018.
92. Gavan J. Fitzsimons and Sarah G. Moore, "Should We Ask Our Children About Sex, Drugs and Rock & Roll? Potentially Harmful Effects of Asking Questions About Risky Behavior," *Journal of Consumer Psychology* 18, no. 2 (Apr 2008): 82–95, https://doi.org/10.1016/j.jcps.2008.01.002.
93. Gavan J. Fitzsimons, Joseph C. Nunes, and Patti Williams, "License to Sin: The Liberating Role of Reporting Expectations," *Journal of Consumer Research* 34, no. 1 (Jun 2007): 22–31, https://doi.org/10.1086/513043.
94. Fitzsimons, Nunes, and Williams, "License to Sin."
95. Fitzsimons, Nunes, and Williams, "License to Sin."
96. Fitzsimons, Nunes, and Williams, "License to Sin."
97. Fitzsimons, Nunes, and Williams, "License to Sin."
98. John Stuart Mill, *On Liberty* (London: John W. Parker and Son, 1859), 277, 292; Candace Cummins Gauthier, "Philosophical Foundations of Respect for Autonomy," *Kennedy Institute of Ethics Journal* 3, no. 1 (1993): 27, https://doi.org/10.1353/ken.0.0103.
99. United Nations, General Assembly, *Interim Report of the Special Rapporteur on Freedom of Religion or Belief, Ahmed Shaheed*, A/76/380, (October 5, 2021), available from https://undocs.org/en/A/76/380.
100. Sofia Moutinho, "Advertisers Could Come for Your Dreams, Researchers Warn," *Science* 372, no. 6549 (2021): 1380, https://doi.org/10.1126/science.372.6549.1380.

101. "Coors Light and Coors Seltzer Are Creating the First Big Game Ad That Runs in Your Dreams," *Business Wire*, January 27, 2021, https://www.businesswire.com/news/home/20210127005208/en/Coors-Light-and-Coors-Seltzer-Are-Creating-the-First-Big-Game-Ad-That-Runs-in-Your-Dreams.
102. Adam Haar Horowitz et al., "Dormio: A Targeted Dream Incubation Device," *Consciousness and Cognition* 83 (Aug 2020): 102938, https://doi.org/10.1016/j.concog.2020.102938.
103. Ruth Cassidy, "Advertisers are After Our Dreams Now, Because We Live in a Nightmare World," *PC Gamer* (July 10, 2021), https://www.pcgamer.com/advertisers-are-after-our-dreams-now-because-we-live-in-a-nightmare-world/.
104. Sofia Moutinho, "Advertisers Could Come for Your Dreams, Researchers Warn," *Science* 372, no. 6549 (2021): 1380, https://doi.org/10.1126/science.372.6549.1380.
105. AJH Haar, Pattie Maes, and Michelle Carr, "A Dream Engineering Ethic," *Infinite Zero*, PubPub, November 25, 2020, https://00.pubpub.org/pub/83843x5m/release/1.
106. Moutinho, "Advertisers Could Come for Your Dreams, Researchers Warn."
107. Haar, Maes, and Carr, "A Dream Engineering Ethic."
108. Haar, Maes, and Carr, "A Dream Engineering Ethic."
109. Moutinho, "Advertisers Could Come for Your Dreams, Researchers Warn."
110. Moutinho, "Advertisers Could Come for Your Dreams, Researchers Warn."
111. Moutinho, "Advertisers Could Come for Your Dreams, Researchers Warn."

Chapter 8: Bewilderbeasts

1. Transcript of Meeting 5, Session 6 of the Presidential Commission for the Study of Bioethical Issues, New York, NY, May 18, 2011, https://bioethicsarchive.georgetown.edu/pcsbi/node/225.html.
2. Transcript of Meeting 5, Session 6 of the Presidential Commission for the Study of Bioethical Issues.
3. Amelia Tait, "'Am I Going Crazy or Am I Being Stalked?' Inside the Disturbing Online World of Gangstalking," *MIT Technology Review*, August 7, 2020, https://www.technologyreview.com/2020/08/07/1006109/inside-gangstalking-disturbing-online-world/.
4. Lorraine P. Sheridan and David V. James, "Complaints of Group-Stalking ('gang-stalking'): An Exploratory Study of their Nature and Impact on Complainants," *The Journal of Forensic Psychiatry & Psychology* 26, no. 5 (2015): 601–623, http://dx.doi.org/10.1080/14789949.2015.1054857.
5. Sheridan and James, "Complaints of Group-Stalking."
6. Amelia Tait, "'Am I Going Crazy or Am I Being Stalked?'"
7. "The Baton Rouge Gunman and 'Targeted Individuals,'" *New York Times* July 19, 2016, https://www.nytimes.com/2016/07/20/us/gavin-long-baton-rouge-targeted-individuals.html.
8. Andrew Lustig, Gavin Brookes, and Daniel Hunt, "Social Semiotics of Gangstalking Evidence Videos on YouTube: Multimodal Discourse Analysis of a Novel Persecutory Belief System," *JMIR Mental Health* 8, no. 10 (Oct 2021): e30311, https://doi.org/10.2196/30311.
9. Tait, "'Am I Going Crazy or Am I Being Stalked?'"; Allen McDuffee, "Conspiracy Theories Abound After Navy Yard Shooting," *Wired*, September 20, 2013, https://www.wired.com/2013/09/navy-yard-conspiracies/.
10. "The Baton Rouge Gunman and 'Targeted Individuals,'" *New York Times*.
11. "FSU Shooter Myron May Feared Government Was Targeting Him: Cops," *NBC News*, November 20, 2014, https://www.nbcnews.com/news/us-news/fsu-shooter-myron-may-feared-government-was-targeting-him-cops-n252731; Mark Schlueb and

Stephen Hudak, "FSU Shooter Myron May Feared 'Energy Weapon', Heard Voices, Thought Police Were Watching Him," *Orlando Sentinel*, November 21, 2014, https://www.orlandosentinel.com/news/breaking-news/os-fsu-shooting-myron-may-update-20141121-story.html.

12. Gerard Albert III, "Gardens Homicide Puts 'Gang Stalking' in Spotlight," *Palm Beach Post*, April 12, 2022, A1.
13. The Associated Press, "Delusional Man Found Competent for Trial in Teen's Stabbing," *ABC News*, June 10, 2022, https://abcnews.go.com/US/wireStory/delusional-man-found-competent-trial-teens-stabbing-85308104.
14. "The Baton Rouge Gunman and 'Targeted Individuals,'" *New York Times*; Kim Whiting, "Government Guinea Pigs?: Investigating the Claims of 'Targeted Individuals' Who Insist They're Being Stalked, Tortured," *The Reporters, Inc.* June 2021, accessed July 14, 2021, https://thereporters.org/letter/government-guinea-pigs/.
15. Brianna Nofil, "The CIA's Appalling Human Experiments with Mind Control," *The History Channel*, accessed July 14, 2022, https://www.history.com/mkultra-operation-midnight-climax-cia-lsd-experiments.
16. Terry Gross, "The CIA's Secret Quest For Mind Control: Torture, LSD And A 'Poisoner In Chief'," *Fresh Air*, NPR, September 9, 2019, https://www.npr.org/2019/09/09/758989641/the-cias-secret-quest-for-mind-control-torture-lsd-and-a-poisoner-in-chief.
17. Nofil, "The CIA's Appalling Human Experiments with Mind Control."
18. Adrian Hartrick and Dominika Ożyńska, "MK-Ultra: The CIA's Secret Pursuit of 'Mind Control'," *BBC Reel* video, March 29, 2022, https://www.bbc.com/reel/video/p0by2ybb/mk-ultra-the-cia-s-secret-pursuit-of-mind-control-.
19. Thomas Hobbs, "The Conspiracy Theorists Convinced Celebrities are Under Mind Control," *Wired*, September 5, 2019, https://www.wired.co.uk/article/mkultra-conspiracy-theory-meme.
20. Nofil, "The CIA's Appalling Human Experiments with Mind Control"; Julie Vanderperre, "Declassified: Mind Control at McGill," *McGill Tribune*, February 2, 2016, https://www.mcgilltribune.com/mind-control-mcgill-mk-ultra/.
21. Gross, "The CIA's Secret Quest For Mind Control: Torture, LSD And A 'Poisoner In Chief.'"
22. Gross, "The CIA's Secret Quest For Mind Control: Torture, LSD And A 'Poisoner In Chief'"; Nofil, "The CIA's Appalling Human Experiments with Mind Control."
23. Vanderperre, "Declassified: Mind Control at McGill."
24. Adrian Hartrick and Dominika Ożyńska, "MK-Ultra: The CIA's Secret Pursuit of 'Mind Control.'" *BBC Reel* video, March 29, 2022, https://www.bbc.com/reel/video/p0by2ybb/mk-ultra-the-cia-s-secret-pursuit-of-mind-control-.
25. Terry Gross, "The CIA's Secret Quest for Mind Control: Torture, LSD and A 'Poisoner In Chief.'"
26. Hartrick and Ożyńska, "MK-Ultra: The CIA's Secret Pursuit of 'Mind Control.'"
27. Nofil, "The CIA's Appalling Human Experiments with Mind Control."
28. Gross, "The CIA's Secret Quest for Mind Control: Torture, LSD and A 'Poisoner In Chief.'"
29. Hobbs, "The Conspiracy Theorists Convinced Celebrities are Under Mind Control."
30. Hartrick and Ożyńska, "MK-Ultra: The CIA's Secret Pursuit of 'Mind Control.'"
31. Nofil, "The CIA's Appalling Human Experiments with Mind Control."
32. Gross, "The CIA's Secret Quest for Mind Control: Torture, LSD and A 'Poisoner In Chief'"; Vanderperre, "Declassified: Mind Control at McGill."
33. Hobbs, "The Conspiracy Theorists Convinced Celebrities are Under Mind Control."
34. Gregg Kilday, "First-Look Trailer: Errol Morris Explores CIA's Secret LSD Experiments in

Netflix Doc," *Hollywood Reporter*, August 28, 2017, https://www.hollywoodreporter.com/movies/movie-news/first-look-trailer-errol-morris-explores-cias-secret-lsd-experiments-netflix-doc-1033144/.
35. Hobbs, "The Conspiracy Theorists Convinced Celebrities are Under Mind Control."
36. Hobbs, "The Conspiracy Theorists Convinced Celebrities are Under Mind Control."
37. Hobbs, "The Conspiracy Theorists Convinced Celebrities are Under Mind Control."
38. Stavros Atlamazoglou, "Warnings About 'Brain-Control' Weapons Reflect Growing US Concern About China's Military Research," *Business Insider*, February 14, 2022, https://www.businessinsider.com/brain-control-weapon-warnings-show-concern-for-china-military-research-2022–2.
39. Nathan Beauchamp-Mustafaga, "Cognitive Domain Operations: The PLA's New Holistic Concept for Influence Operations," *China Brief* 19, no. 16 (Sep 6, 2019), https://jamestown.org/program/cognitive-domain-operations-the-plas-new-holistic-concept-for-influence-operations/.
40. Office of the Secretary of Defense, *Annual Report to Congress: Military and Security Developments Involving the People's Republic of China, 2021*, November 3, 2021, available at https://media.defense.gov/2021/Nov/03/2002885874/-1/-1/0/2021-CMPR-FINAL.PDF.
41. Office of the Secretary of Defense, *Annual Report to Congress: Military and Security Developments Involving the People's Republic of China.*
42. Beauchamp-Mustafaga, "Cognitive Domain Operations: The PLA's New Holistic Concept for Influence Operations."
43. Beauchamp-Mustafaga, "Cognitive Domain Operations."
44. Bill Gertz, "Chinese 'Brain Control' Warfare Work Revealed," *Washington Times*, December 29, 2021, https://www.washingtontimes.com/news/2021/dec/29/pla-brain-control-warfare-work-revealed/.
45. Hai Jin, Li-Jun Hou, and Zheng-Guo Wang, "Military Brain Science—How to Influence Future Wars," *Chinese Journal of Traumatology* (in English) 21, no. 5 (2018): 277–80, https://doi.org/10.1016/j.cjtee.2018.01.006. https://www.sciencedirect.com/science/article/pii/S1008127517303188.
46. Jin, Hou, and Wang, "Military Brain Science."
47. Jin, Hou, and Wang, "Military Brain Science."
48. Anika Binnendijk, Timothy Marler, and Elizabeth M. Bartels, *Brain-Computer Interfaces: U.S. Military Applications and Implications, An Initial Assessment* (Santa Monica, CA: RAND Corporation, 2020), https://www.rand.org/pubs/research_reports/RR2996.html.
49. Zhang Huang and Jia Zhenzhen, "Human–Machine Integration in Unmanned Combat: Challenges and Solutions," *National Defense Science & Technology* 41, no. 6 (2020): 105–109, https://caod.oriprobe.com/articles/60604661/Human_machine_integration_in_unmanned_combat__chal.htm.
50. Mark Hodge, "Inside China's Terrifying 'Brain Control Weapons' Capable of 'Paralyzing Enemies,'" *New York Post*, December 31, 2021, https://nypost.com/2021/12/31/inside-chinas-terrifying-brain-control-weapons-capable-of-paralyzing-enemies/.
51. Joe Saballa, "US: China Developing 'Brain Control Weaponry,'" *The Defense Post*, December 20, 2021, https://www.thedefensepost.com/2021/12/20/us-china-brain-control-weaponry/.
52. Scott Simon, "'The Neuroscientist Who Lost Her Mind' Returns From Madness," *NPR*, March 31, 2018, https://www.npr.org/sections/health-shots/2018/03/31/598236622/the-neuroscientist-who-lost-her-mind-returns-from-madness.
53. Benedict Carey, "Were U.S. Diplomats Attacked in Cuba? Brain Study Deepens Mystery," *New York Times*, July 23, 2019, https://www.nytimes.com/2019/07/23/science/cuba-diplomats-health.html.

54. Scott Pelley, "Havana Syndrome: High-Level National Security Officials Stricken With Unexplained Illness on White House Grounds," *60 Minutes*, CBS News, February 20, 2022, https://www.cbsnews.com/news/havana-syndrome-white-house-cabinet-60-minutes-2022–02–20/.
55. Scott Pelley, "Havana Syndrome.": High-Level National Security Officials Stricken With Unexplained Illness on White House Grounds," *60 Minutes*, CBS News, February 20, 2022, https://www.cbsnews.com/news/havana-syndrome-white-house-cabinet-60-minutes-2022–02–20/.
56. National Academies of Sciences, Engineering, and Medicine, *An Assessment of Illness in U.S. Government Employees and Their Families at Overseas Embassies*, ed. David A. Relman and Julie A. Pavlin, (Washington, DC, The National Academies Press, 2020), 11.
57. Julian Borger, "US Officials Confirm 130 Incidents of Mysterious Havana Syndrome Brain Injury," *Guardian*, May 13, 2021, https://www.theguardian.com/us-news/2021/may/13/havana-syndrome-brain-injury-130-incidents.
58. Jean Guerrero, "Are Electromagnetic Weapons Involved? Taking Victims of 'Havana Syndrome' Seriously," *Los Angeles Times*, February 24, 2022, https://www.latimes.com/opinion/story/2022–02–24/electromagnetic-weapons-havana-syndrome.
59. Olivia Gazis, "CIA 'Havana Syndrome' Task Force Rules Out Foreign Attack in Most Reported Cases in Interim Report, " *CBS News*, January 20, 2022, https://www.cbsnews.com/news/havana-syndrome-cia-task-force-foreign-attacks-rule-out/.
60. National Academies of Sciences, Engineering, and Medicine, *An Assessment of Illness in U.S. Government Employees and Their Families at Overseas Embassies*, ed. David A. Relman and Julie A. Pavlin, (Washington, DC, The National Academies Press, 2020), 1.
61. Ragini Verma et al., "Neuroimaging Findings in US Government Personnel with Possible Exposure to Directional Phenomena in Havana, Cuba," *JAMA* 322, no. 4 (2019): 336–347, https://doi.org/10.1001/jama.2019.9269.
62. National Academies of Sciences, Engineering, and Medicine, *An Assessment of Illness in U.S. Government Employees and Their Families at Overseas Embassies*.
63. Cheryl Rofer, "Claims of Microwave Attacks are Scientifically Implausible," *Foreign Policy*, May 10, 2021, https://foreignpolicy.com/2021/05/10/microwave-attacks-havana-syndrome-scientifically-implausible/.
64. Tim Moore, "For Your Ears Only: What's Really Behind Havana Syndrome," *Sydney Morning Herald*, May 27, 2022, https://www.smh.com.au/national/for-your-ears-only-what-s-really-behind-havana-syndrome-20220506-p5aj70.html.
65. Moore, "For Your Ears Only."
66. Moore, "For Your Ears Only."
67. Office of the Director of National Intelligence, "Statement from DNI Haines and DCIA Burnes," Press Release No. 1–22, February 2, 2022, https://www.dni.gov/index.php/newsroom/press-releases/press-releases-2022/item/2274-statement-from-dni-haines-and-dcia-burns.
68. Olivia Gazis, "Expert Panel Reaffirms Directed Energy Could Be Behind 'Havana Syndrome' Cases," *CBS News* (February 3, 2022), https://www.cbsnews.com/news/havana-syndrome-cases-directed-energy/.
69. Office of the Director of National Intelligence, "Statement from DNI Haines and DCIA Burnes."
70. Gazis, "Expert Panel Reaffirms Directed Energy Could Be Behind 'Havana Syndrome.'"
71. Andrew Desiderio, "Lawmakers Skewer Interim CIA Report on Havana Syndrome," *Politico*, January 21, 2022.
72. Julian Borger, "Microwave Weapons That Could Cause Havana Syndrome Exist, Experts Say," *The Guardian*, June 2, 2021, https://www.theguardian.com/science/2021/jun/02/microwave-weapons-havana-syndrome-experts.

73. The White House, "Background Press Call By Senior Administration Officials On New Cuba Policy," press briefing via teleconference, May 17, 2022, https://www.whitehouse.gov/briefing-room/press-briefings/2022/05/17/background-press-call-by-senior-administration-officials-on-new-cuba-policy/.
74. François du Cluzel, *Cognitive Warfare*, NATO Innovation Hub, November 2020, 20, https://www.innovationhub-act.org/sites/default/files/2021–01/20210122_CW%20Final.pdf.
75. François du Cluzel, *Cognitive Warfare*, p. 31.
76. General Assembly Resolution 39/46: *Convention Against Torture and Other Cruel, Inhuman or Degrading Treatment or Punishment*, A/RES/39/46 (December 10, 1984), available from https://undocs.org/en/A/RES/39/46.
77. United Nations, General Assembly, *Torture and Other Cruel, Inhuman or Degrading Treatment or Punishment: Report of the Special Rapporteur*, A/HRC/43/49 (March 20, 2020), available from https://undocs.org/en/A/HRC/43/49.
78. General Assembly Resolution 39/46: *Convention Against Torture and Other Cruel, Inhuman or Degrading Treatment or Punishment*, A/RES/39/46 (December 10, 1984), available from https://undocs.org/en/A/RES/39/46.
79. United Nations, General Assembly, *Torture and Other Cruel, Inhuman or Degrading Treatment or Punishment: Report of the Special Rapporteur.*
80. United Nations, General Assembly, *Torture and Other Cruel, Inhuman or Degrading Treatment or Punishment: Report of the Special Rapporteur.*

Chapter 9: Beyond Human

1. Zvonimir Vrselja et al., "Restoration of Brain Circulation and Cellular Functions Hours Post-Mortem," *Nature* 568 (2019): 336–343, https://doi.org/10.1038/s41586-019-1099–1.
2. Nita A. Farahany, Henry T. Greely and Charles Giattino, "Part-Revived Pig Brains Raise Slew of Ethical Quandaries," *Nature*, April 17, 2019, https://www.nature.com/articles/d41586-019-01168–9.
3. Benjamin D. Ross, "Transhumanism: An Ontology of the World's Most Dangerous Idea" (PhD dissertation, University of North Texas, May 2019), https://digital.library.unt.edu/ark:/67531/metadc1505282/.
4. Celina Ribeiro, "Beyond Our 'Ape-Brained Meat Sacks': Can Transhumanism Save Our Species?," *The Guardian*, June 3, 2022, https://www.theguardian.com/books/2022/jun/04/beyond-our-ape-brained-meat-sacks-can-transhumanism-save-our-species.
5. Ribeiro, "Beyond Our 'Ape-Brained Meat Sacks.'"
6. Ribeiro.
7. Raffi Khatchadourian, "The Doomsday Invention: Will Artificial Intelligence Bring Us Utopia or Destruction?," *New Yorker*, November 23, 2015, https://www.newyorker.com/magazine/2015/11/23/doomsday-invention-artificial-intelligence-nick-bostrom.
8. Nick Bostrom, "A History of Transhumanist Thought," *Journal of Evolution and Technology* 14, no. 1 (April 2005).
9. Khatchadourian, "The Doomsday Invention."
10. Bernard Marr, "Is Artificial Intelligence (AI) a Threat to Humans?" *Forbes,* March 2, 2020, https://www.forbes.com/sites/bernardmarr/2020/03/02/is-artificial-intelligence-ai-a-threat-to-humans/?sh=7445a95f205d.
11. Ross, "Transhumanism."
12. Ross, "Transhumanism."
13. Bostrom, "History of Transhumanist Thought."
14. Francis Fukuyama, "Transhumanism," *Foreign Policy* 144 (Sept/Oct 2004): 42–43.

15. Francis Fukuyama, *Our Posthuman Future: Consequences of the Biotechnology Revolution* (New York: Farrar, Straus and Giroux, 2002).
16. Michael J. Sandel, *The Case Against Perfection: Ethics in the Age of Genetic Engineering* (Cambridge, MA: Harvard University Press, 2009), 26–27.
17. Ross, "Transhumanism: An Ontology of the World's Most Dangerous Idea."
18. Calico, "Research & Technology," Calico Life Sciences LLC, accessed July 14, 2022, https://www.calicolabs.com/research-technology.
19. Maya Kosoff, "Peter Thiel Wants to Inject Himself with Young People's Blood," *Vanity Fair*, August 1, 2016, https://www.vanityfair.com/news/2016/08/peter-thiel-wants-to -inject-himself-with-young-peoples-blood.
20. Alcor, "Membership," Alcor Life Extension Foundation, accessed July 14, 2022, https://www.alcor.org/membership/.
21. Khatchadourian, "The Doomsday Invention."
22. Antonio Regalado, "A Startup is Pitching a Mind-Uploading Service that is '100 percent fatal,'" *MIT Technology Review*, March 13, 2018, https://www.technologyreview .com/2018/03/13/144721/a-startup-is-pitching-a-mind-uploading-service-that-is -100-percent-fatal/.
23. Regalado, "A Startup is Pitching a Mind-Uploading Service that is '100 percent fatal.'"
24. Ujwal Chaudhary et al., "Spelling Interface Using Intracortical Signals in a Completely Locked-in Patient Enabled Via Auditory Neurofeedback Training," *Nature Communications* 13, no. 1236 (2022), https://doi.org/10.1038/s41467-022-28859–8.
25. Gemma Ross, "Paralysed Man Says First Words in Months with Brain Implant: 'I Want a Beer,'" *MixMag*, March 25, 2022, https://mixmag.net/read/paralysed man -says-first-words-months-brain-implant-want-beer-news.
26. "Meet a Pioneer in Stroke Recovery," *CBS News*, produced by Amiel Weisfogel, January 31, 2021, https://www.cbsnews.com/news/meet-a-pioneer-in-stroke-recovery/.
27. "Moonshot: Our Mission Matters," Cognixion, accessed July 12, 2022, https://www .cognixion.com/#moonshot-section.
28. Isobel Asher Hamilton, "The Story of Neuralink: Elon Musk's AI Brain Chip Company Where he had Twins with a Top Executive," *Business Insider*, via Yahoo News, July 7, 2022, https://uk.news.yahoo.com/news/everything-know-neuralink-elon -musks-103000152.html.
29. Will Nicol, "Elon Musk Says Neuralink Will Be Like a 'Fitbit in Your Skull with Tiny Wires," *Digital Trends*, August 29, 2020, https://www.digitaltrends.com/news/neuralink -progress-update-2020/.
30. Hamilton, "The Story of Neuralink."
31. Mass General Press Office, "First Evidence of Replay During Sleep in the Human Motor Cortex, Which Governs Voluntary Movement," *Neuroscience News*, June 25, 2022, https://neurosciencenews.com/motor-replay-sleep-20906/.
32. Anders Sandberg, "Feasibility of Whole Brain Emulation," *Philosophy and Theory of Artificial Intelligence*, vol 5, ed. Vincent C. Müller, (Berlin/Heidelberg: Springer, 2013), 251–264, https://doi.org/10.1007/978-3-642-31674-6_19.
33. Hamilton, "The Story of Neuralink."
34. Mark Sullivan, "Synchron's Nonsurgically Implanted BCI Could Offer New Hope for Paraplegics," *Fast Company*, June 3, 2022, https://www.fastcompany.com/90756530 /synchrons-nonsurgically-implanted-bci-could-offer-new-hope-for-paraplegics.
35. Sullivan, "Synchron's Nonsurgically Implanted BCI Could Offer New Hope for Paraplegics."
36. "Second Sight Medical Products—Life in a New Light," Second Sight, accessed July 14, 2022, https://secondsight.com.
37. Michelle Z. Donahue, "How a Color-Blind Artist Became the World's First Cyborg,"

National Geographic, April 3, 2017, https://www.nationalgeographic.com/science/article/worlds-first-cyborg-human-evolution-science.

38. Sarah McQuate-Washington, "With BrainNet, 3 People Play Tetris with Their Minds," *Futurity*, July 2, 2019, https://www.futurity.org/brainnet-brain-to-brain-interface-game-2098002/.
39. McQuate-Washington, "With BrainNet, 3 People Play Tetris with Their Minds."
40. McQuate-Washington, "With BrainNet, 3 People Play Tetris with Their Minds."
41. Qian Ma, *Directly Wireless Communication of Human Minds Via Non-Invasive Brain-Computer Metasurface Platform*, eLight 2, 11 (2022). https://doi.org/10.1186/s43593-022-00019-x.
42. George K. Jaravas, *A Hybrid BMI for Control of Robotic Swarms: Preliminary Results*, 2017 IEEE/RSJ International Conference on Intelligent Robots and Systems (IROS) (September 24–28, 2017), https://cpb-us-w2.wpmucdn.com/sites.udel.edu/dist/b/9405/files/2019/08/IROS17_Karavas.pdf.
43. Maggie Fox, "Brain Stimulation Acts 'Like a Switch' to Turn off Severe Depression for One Patient," *CNN*, October 4, 2021, https://www.cnn.com/2021/10/04/health/depression-implant-treatment-wellness/index.html.
44. Fox, "Brain Stimulation Acts 'Like a Switch' to Turn off Severe Depression for One Patient."
45. Halo Neuroscience, "Olympians Train with Halo Sport to Prepare for Rio 2016," *PR Newswire*, July 26, 2016, https://www.prnewswire.com/news-releases/olympians-train-with-halo-sport-to-prepare-for-rio-2016–300303734.html.
46. Homepage, Flow Neuroscience, accessed July 14, 2022, https://www.flowneuroscience.com.
47. Joelle Renstrom, "It's the End of the World as We Know it and We Feel Fantastic: Examining the End of Suffering," *NANO* 13 (Dec 2018), https://nanocrit.com/issues/issue13/It-s-the-End-of-the-World-as-We-Know-It-and-We-Feel-Fantastic-Examining-the-End-of-Suffering.
48. Renstrom, "We Feel Fantastic."
49. Emily R. D. Murphy, "Collective Cognitive Capital," *William & Mary Law Review* 63, no. 4 (Mar 2022): 1347–1408, https://heinonline.org/HOL/P?h=hein.journals/wmlr63&i=1387.
50. Yizhi Liu, Mahmoud Habibnezhad, and Houtan Jebelli, "Brainwave-Driven Human-Robot Collaboration in Construction," *Automation in Construction* 124 (Apr 2021), https://doi.org/10.1016/j.autcon.2021.103556.
51. "These Robots Want to Read Your Mind While You Work—You Should Let Them," Inverse.com, https://www.inverse.com/innovation/mind-reading-robots-are-the-future-of-work.
52. "Research Proves Your Brain Needs Breaks," *WorkLab*, Microsoft, April 20, 2021, https://www.microsoft.com/en-us/worklab/work-trend-index/brain-research.
53. Sarah Cascone, "Your Brain May Not be Able to Distinguish a Digital Reproduction of an Artwork from the Real Thing, A New Study Suggests," *Artnet*, June 10, 2020, https://news.artnet.com/art-world/brain-digital-art-reproduction-study-1873623.
54. Michael Vincent, "Yo-Yo Ma Surprises Montréal With Free Performance In The Métro (Video)," *Ludwig Van Toronto*, December 9, 2018, https://www.ludwig-van.com/toronto/2018/12/09/the-scoop-yo-yo-ma-surprises-montreal-with-free-performance-in-the-metro-video/.
55. Lisa Park, "Eunoia" (video performance), 2013, https://www.thelisapark.com/work/eunoia.
56. Mark Wilson, "A Machine that Turns Brain Waves into Music," *Fast Company*, March 9, 2012, https://www.fastcompany.com/90186468/a-machine-that-turns-brain-waves-into-music.

57. Michael Specter, "Sense and Synthetic Biology," *New Yorker*, December 16, 2010, https://www.newyorker.com/news/news-desk/sense-and-synthetic-biology.
58. Barack Obama to Amy Gutmann, Letter to the Chair of the Presidential Commission on the Study of Bioethical Issues, July 1, 2013, in *Gray Matters: Topics at the Intersection of Neuroscience, Ethics, and Society* 2 (Mar 2015): vi-vii. https://bioethicsarchive.georgetown.edu/pcsbi/sites/default/files/news/Letter-from-President-Obama-05.20.10.pdf.
59. Presidential Commission for the Study of Bioethical Issues, *New Directions: The Ethics of Synthetic Biology and Emerging Technologies*, (Washington, DC: GPO, December 2010), available from https://bioethicsarchive.georgetown.edu/pcsbi/sites/default/files/PCSBI-Synthetic-Biology-Report-12.16.10_0.pdf.
60. Presidential Commission for the Study of Bioethical Issues, *New Directions*.
61. John Donnelly, "Aristotle Would be Proud: 'Prudent Vigilance' for Synthetic Biology," The Blog of the 2009–2017 Presidential Commission for the Study of Bioethical Issues, November 16, 2010, https://bioethicsarchive.georgetown.edu/pcsbi/blog/2010/11/16/aristotle-would-be-proud-prudent-vigilance'-for-synthetic-biology/index.html.
62. Henry T. Greely, "CRISPR'd Babies: Human Germline Genome Editing in the 'He Jiankui Affair,'" *Journal of Law and the Biosciences*, 6. 111–83 (October 2019).
63. Greely, "CRISPR'd Babies."
64. Greely, "CRISPR'd Babies."
65. Greely, "CRISPR'd Babies."
66. Greely, "CRISPR'd Babies."

Chapter 10: On Cognitive Liberty

1. Brandon L. Garrett, Laurence R. Helfer, and Jayne C. Huckerby, "Closing International Law's Innocence Gap," *Southern California Law Review* 95 (2021): 311–64, 328.
2. Garrett, Helfer, and Huckerby, "Closing International Law's Innocence Gap," 328–31.
3. Garrett, Helfer, and Huckerby, "Closing International Law's Innocence Gap," 328.
4. Garrett, Helfer, and Huckerby, "Closing International Law's Innocence Gap," 329–30.
5. Garrett, Helfer, and Huckerby, "Closing International Law's Innocence Gap," 333.

Afterword

1. "An Interface Connects," *Nature Electronics* 6, no. 2 (February 28, 2023): 89; Eisenstein, Michael Eisenstein, "Seven Technologies to Watch in 2024," *Nature* 625 (2024): 848.
2. Henri Lorach et al., "Walking Naturally after Spinal Cord Injury Using a Brain–Spine Interface," *Nature* 618, no. 7963 (May 24, 2023): 126–33, https://doi.org/10.1038/s41586-023-06094-5.
3. Musk, Elon (@elonmusk), "The first human received an implant from @Neuralink yesterday and is recovering well. Initial results show promising neuron spike detection," X, January 29, 2024, https://twitter.com/elonmusk/status/1752098683024220632.
4. Musk, Elon (@elonmusk), "Enables control of your phone or computer, and through them almost any device, just by thinking. Initial users will be those who have lost the use of their limbs. Imagine if Stephen Hawking could communicate faster than a speed typist or auctioneer. That is the goal," X, January 29, 2024, https://twitter.com/elonmusk/status/1752119586470949056.
5. "Home," Neuralink, Accessed January 31, 2024, https://neuralink.com.
6. Jerry Tang et al., "Semantic Reconstruction of Continuous Language from Non-Invasive Brain Recordings," *Nature Neuroscience* 26, no. 5 (May 2023): 858–66, https://doi.org/10.1038/s41593-023-01304-9.
7. "Building the Future with Michael Abrash and Andrew 'Boz' Bosworth: Meta Quest

Blog," Meta, accessed February 16, 2024, https://www.meta.com/blog/quest/building-future-michael-abrash-andrew-boz-bosworth-connect-2023/.

8. Alexandre Défossez et al., "Decoding Speech Perception from Non-Invasive Brain Recordings," *Nature Machine Intelligence* 5, no. 10 (October 5, 2023): 1097–1107, https://doi.org/10.1038/s42256-023-00714-5.
9. Yiqun Duan, et al., "DeWave: Discrete EEG Waves Encoding for Brain Dynamics to Text Translation," preprint, submitted September 25, 2023, last revised January 3, 2024, arXiv:2309.14030v4, https://doi.org/10.48550/arXiv.2309.14030.
10. Erdrin Azemi, et al., "Biosignal Sensing Device Using Dynamic Selection of Electrodes," patent application US20230225659A1, filed January 9, 2023, published July 20, 2023.
11. Conor Russomanno, "A Powerful New Neurotech Tool for Augmenting Your Mind," TED2023, April 2023, https://www.ted.com/talks/conor_russomanno_a_powerful_new_neurotech_tool_for_augmenting_your_mind?language=en.
12. Daniel S. Hain, et al., *Unveiling the Neurotechnology Landscape: Scientific Advancements, Innovations, and Major Trends* (Paris: UNESCO, 2023), https://doi.org/10.54678/OCBM4164.
13. "Apple Provides Powerful Insights into New Areas of Health," Apple, press release, June 5, 2023, https://www.apple.com/newsroom/2023/06/apple-provides-powerful-insights-into-new-areas-of-health/.
14. Daniel S. Hain, et al., *Unveiling the Neurotechnology Landscape: Scientific Advancements, Innovations, and Major Trends.*
15. Information Commissioner's Office, *ICO Tech Futures: Neurotechnology*, June 1, 2023, https://ico.org.uk/media/about-the-ico/research-and-reports/ico-tech-futures-neurotechnology-0-1.pdf.
16. Refik Anadol, *Unsupervised*, Museum of Modern Art (MoMA), November 19, 2022–October 29, 2023, https://www.moma.org/calendar/exhibitions/5535.
17. Neuroelectrics (@Neuroelectrics), "🧠 Here's a glimpse of our project with artist @refikanadol for his exhibition #Unsupervised at #MoMA. Together, we're exploring how the brain responds to #art. We believe that combining art and the latest technology can lead to incredible outcomes. Stay tuned! ✨ #AIandArt," X, September 28, 2023, https://twitter.com/Neuroelectrics/status/1707386746403487868.
18. Sherry Turkle, *Alone Together: Why We Expect More from Technology and Less from Each Other* (New York: Basic Books, 2017).
19. Joseph B. Pine II and James H. Gilmore, "Welcome to the Experience Economy," *Harvard Business Review*, July–August 1998, https://hbr.org/1998/07/welcome-to-the-experience-economy..
20. Noor Al-Sibai, "Grimes Using Custom Brain Interface from Elon Musk Competitor," *The Byte*, March 23, 2023, https://futurism.com/the-byte/grimes-brain-interface-musk-competitor.
21. AJ Keller, LinkedIn post, April 2023, https://www.linkedin.com/posts/andrewjaykeller_neurosity-125-sales-growth-activity-7055196837527388161-Uwv7.
22. Evan Selinger and Judy Hyojoo Rhee, "Normalizing Surveillance," *Northern European Journal of Philosophy* 22, no. 1 (2021): 49–74, https://ssrn.com/abstract=3883551.
23. Melissa Klein and Isabel Vincent, "Albany Doctor Allegedly Ran Sick Human Experiments for Nxivm," *New York Post*, May 5, 2018, https://nypost.com/2018/05/05/nxivm-doctor-charged-with-conducting-human-experiments/.
24. Nicole Chavez and Sonia Moghe, "6 Weeks of Testimony in Nxivm Case Reveal Lurid Details of Alleged Sex Cult, Including Branding Women and Holding Them Captive," CNN, June 16, 2019, https://www.cnn.com/2019/06/16/us/nxivm-keith-raniere-trial/index.html.

25. E. Stoycheff, "Under Surveillance: Examining Facebook's Spiral of Silence Effects in the Wake of NSA Internet Monitoring," *Journalism & Mass Communication Quarterly* 93, no. 2 (2016): 296–311. https://doi.org/10.1177/1077699016630255.
26. Long Qiao, "Chinese Researchers Develop Device They Say Can Test Loyalty of Ruling Party Members," Radio Free Asia, July 4, 2022, https://www.rfa.org/english/news/china/polygraph-loyalty-07042022131133.html.
27. Ameya Paleja, "Researchers Develop a 'Mind-Reading' Device to Help Censor Porn in China," *Interesting Engineering,* June 21, 2022, https://interestingengineering.com/culture/mind-reading-device-to-censor-porn.
28. Nita Farhany, "Cultivating Cognitive Liberty in the Age of Generative AI," *Microsoft Unlocked,* June 19, 2023, https://unlocked.microsoft.com/ai-anthology/nita-farahany/.
29. Hunter T. Carter, "Neural Rights: Landmark Ruling," ArentFox Schiff, October 18, 2023, https://www.afslaw.com/perspectives/news/neural-rights-landmark-ruling.
30. Lex Fridman, "Transcript for Mark Zuckerberg: First Interview in the Metaverse," *Lex Fridman Podcast,* June 19, 2023, https://lexfridman.com/mark-zuckerberg-3-transcript/.
31. Mike Isaac, "Zuckerberg turns to face abuse victim families: 'I'm sorry for everything you have all been through,'" *New York Times,* January 31, 2024, https://www.nytimes.com/live/2024/01/31/technology/child-safety-senate hearing.
32. Jim Henry, "EV Appeal Growing as Risk and Novelty Fade," *Forbes,* September 26, 2023, https://www.forbes.com/sites/jimhenry/2023/09/26/ev-appeal-growing-as-risk-and-novelty-fade/?sh=97980a66ba69.

Index

About the Author

Merritt Chesson

Nita Farahany is the Robinson O. Everett Distinguished Professor of Law & Philosophy at Duke University and founding director of the Duke Initiative for Science & Society. She is a frequent commentator for national media and radio and keynote speaker at events including TED, the Aspen Ideas Festival, the World Economic Forum, and judicial conferences worldwide. From 2010–2017, she served on the U.S. Presidential Commission for the Study of Bioethical Issues. She is also co–editor in chief of the *Journal of Law and the Biosciences* and is on the board of advisors for *Scientific American.* Farahany holds an AB in genetics from Dartmouth College, an ALM in biology from Harvard University, and a JD, MA, and PhD in philosophy from Duke University.